I0022200

Global Climate Change and Natural Resources 2014

Also by Emil Morhardt

Global Climate Change and Natural Resources 2013
Global Climate Change and Natural Resources 2012
Ecological Consequences of Climate Change 2012
Global Climate Change and Natural Resources 2011
Ecological Consequences of Global Change 2011
Climate Change and Natural Resources 2010
Ecological Consequences of Global Climate Change: Summaries of the 2009 Scientific Literature
Global Climate Change and Natural Resources: Summaries of the 2007–2008 Scientific Literature
Biology of Global Change
Global Climate Change: Summaries of the 2006–2007 Scientific Literature
Research in Natural Resources Management
Clean Green and Read All Over
Research in Ecosystem Services
California Desert Flowers
Cannon and Slinkard Fire Recovery Study: A Photographic Flora

Global Climate Change and Natural Resources 2014

J. Emil Morhardt, Editor

Claremont Academic Press

Claremont Academic Press
W. M. Keck Science Department
925 N. Mills Avenue
Claremont, California 91711
(909) 621-8190

Morhardt, J. Emil
Global Climate Change and Natural Resources 2014/ J. Emil Morhardt, Editor.

ISBN 978-0-9843823-8-5 (paper)

Table of Contents

Forward

J. Emil Morhardt

Earlier this year I attended a departmental seminar (here in the W. M. Keck Science Department at the Claremont Colleges) in which the speaker, a well-known expert on global climate change advised us never to use the term "global warming". He said that it was too confusing since a good bit of the time it looks like we are getting more, not less snow and ice, and that it hasn't been that warm in Southern California lately. He said that we should always refer to "climate change" instead. But he and we know that the cause of this climate change is global warming, whether it feels warmer to us or not. It is the increased heat retention by the Earth, caused primarily by the carbon dioxide released from burning fossil fuels, and secondarily by methane released from agricultural production that is the direct cause of climate change, and no amount of dancing around the subject is going to alter that fact.

In any case, it turns out that whether people who identified themselves as politically independent—most of us and probably the majority of scientists—believe in climate change has to do with what the weather is doing right now. As reported by Lawrence Hamilton and Mary Stampone in the journal *Weather, Climate, and Society* (2013), if "Interviewed on unseasonably warm days, independents tend to agree with the scientific consensus regarding anthropogenic climate change. On unseasonably cool days, they tend not to agree. Although temperature effects are sharpest for just a 2-day window, positive effects are seen for longer windows as well. As future climate change shifts the distribution of anomalies and extremes, this will first affect beliefs among unaligned voters." Conservatives were not swayed and uniformly reject the concept of climate change. Liberals, the opposite, tending to go along with the majority of scientists.

This politicization of scientific beliefs is not new. Conservatives, by definition, don't like change, and apparently don't want to acknowledge it even if it has little to do with the political will. Consider how long it took the Catholic Church to come to grips with the fact that the Earth revolves around the Sun, rather than vice versa. It isn't clear to me why liberals tend to believe in the scientific consensus since they tend to resist being told what to believe as well, but it is good that at least one identifiable segment of society is willing to deal with it.

As you will discover reading this book, there is very little scientific debate about whether anthropogenic greenhouse gas releases are warming the planet, but there is vast uncertainty about when, how much, and where the climate will change, and even more uncertainty about the consequences for sea level, agriculture, ecology, and human well being. We are at the beginning of a centuries-long change in the Earth's climate, and resulting effects could come suddenly or slowly—we won't know until they happen. I hope that slowly is in fact how they occur, but there are sufficient records of rapid climate change in the past, that we should be prepared to act if we need to. As of now, we are completely unprepared.

References Cited

Hamilton, L.C., Stampone, M.D., 2013. Blowin'in the wind: Short-term weather and belief in anthropogenic climate change. Weather, Climate, and Society 5, 112–119.

Section I: Energy Issues

1. Environmental Impacts of Shale Gas Development

Shannon Julius

Shale gas is an unconventional source of natural gas recently made accessible by the technologies of horizontal drilling and hydraulic fracturing. Shale gas and other unconventional sources of natural gas have caused production of methane in the US to increase by 30% since 2005. Despite its increasing importance and potential risks, the environmental implications of producing shale gas have not yet been studied extensively.

The process of extraction is fairly complicated because shale gases are confined within tight shale rock formations deep underground. To reach these shales, a vertical well is drilled up to 8,000 feet underground—below the depth of any conventional natural gas wells—then extended horizontally up to 10,000 feet. The well shaft is contained by a steel casing and sealed off from the surrounding rock by a cylindrical layer of cement. After the well is completed, hydraulic fracturing begins. Hydraulic fracturing involves pumping large quantities of fracturing fluid (water mixed with sand and chemicals) into the well at extremely high pressure. This process creates fractures in the shale formation that extend up and down from the end of the horizontal well. The fractures are kept open by the sand proppants in the fracturing fluid and natural gas is freed to move through the rock formations, into the well, and up to be captured at the surface. Hydraulic fracturing is done many times over the course of a shale gas well's lifetime in order to create fractures along the entire length of the horizontal well. Deep vertical drilling, horizontal drilling, and repeated hydraulic fracturing allow shale gas wells to extract much more natural gas than conventional wells.

Shale formations in the US are extensive, the largest being the Marcellus shale in the Northeast. Taking advantage of these resources in the US has increased the total amount of recoverable natural gas, lowered natural gas prices significantly, and reduced dependence on coal. It is widely believed that natural gas will be the transition fuel from fossil fuels to renewables, thus its development it important for global greenhouse gas reduction. However, the extensive and rapid expansion of shale gas extraction has a number of potential negative consequences for human health, biodiversity, and ecological systems. Of particular concern is the potential contamination of groundwater or surface water from the chemicals present in fracturing fluid and the water that resurfaces after hydraulic fracturing occurs ("wastewater"). Other concerns are natural gas contamination of groundwater via migration from wells and induced seismicity from disposing of wastewater in underground injection wells. These risks have gone largely under-researched and unregulated, as it has been difficult for scientists and lawmakers to keep up with the extensive shale gas development that has occurred since 2008. Recently published research is beginning to investigate the extent of these potential threats.

Review of Impacts of Shale Gas Development on Regional Water Quality

Drilling into shale is a difficult task, as gases are under high pressure and can easily damage the well's integrity if drilling is done incorrectly. Such damage allows natural gases, particularly methane, to "migrate" through cement seals and into groundwater, which happens with approximately 1–3% of wells in Pennsylvania. The high toxicity of fracturing fluid raises the concern of fluid migration accompanying methane migration, and research has yet to determine the extent to which fracturing fluid can affect groundwater. However, it is highly likely that most of the unrecovered fracturing fluid is absorbed by the shale formation. The remaining fracturing fluid is recovered as wastewater, which must be disposed of properly in order to maintain surface water quality. These aspects of hydraulic fracturing create environmental and social concerns with respect to regional water quality. Vidic *et al.* (2013) reviewed current research related to chemical aspects of water quality in regions of shale gas well drilling, focusing on the Marcellus shale region. The three major areas of concern were methane migration into groundwater, fracturing fluid contamination of groundwater, and wastewater management.

Vidic *et al.* found reports that indicate that 1–2% of shale gas wells have a faulty cement seal around well casings that are meant to keep methane out of aquifers. This percentage could be slightly higher in actuality, as the Pennsylvania Department of Environmental Protection (PADEP) issued notices of violation to 3.4% of Pennsylvania wells for well construction problems from 2008 to 2013. Damage to well casings can occur when drilling into areas of existing high pressure gases. If drilling pressure is too high, drilling can fracture the formation in undesired places. If pressure is too low, gases can infiltrate the cement seal before it hardens. Either of these possibilities will cause gas to "migrate," to move through the area that has been drilled and potentially end up in nearby groundwater.

Methane alone is not toxic to ingest, but certain aquatic bacteria will remove oxygen from the water if methane is present. Low oxygen conditions can increase the solubility of elements such as arsenic or iron as well as support the life of anaerobic bacteria which create sulfide, an inorganic molecule that can contribute to air and water quality issues. When methane turns returns to a gaseous state after being dissolved, it can cause water turbidity or, in high concentrations, explode.

Dissolved methane levels have been shown to be higher in drinking water wells less than 1 kilometer from active Marcellus gas wells than from wells farther away. However, methane can be formed naturally by subsurface organic matter at high temperatures ("thermogenic" source) or by bacterial processes at shallow depths ("biogenic" source). Shale gas wells extract thermogenic natural gas, so determining whether gas is thermogenic or biogenic is the first step towards figuring out if stray gases may have originated from a shale gas well. With the widespread lack of pre-drilling data, researchers must look closely at groundwater chemistry in order to determine if the source of dissolved gases is thermogenic or biogenic. Methods include measuring the concentrations of other natural gases and looking at the isotopic signatures of hydrogen, oxygen, and carbon. In Pennsylvania, natural methane levels happen to be relatively high, so there is controversy over whether Marcellus wells are to blame for any methane present in groundwater. Compounding the problem is the history of conventional gas and oil drilling; these wells could have been responsible for methane migration in the region before public concern over hydraulic fracturing led to groundwater research.

Fracturing fluid is a potentially hazardous water contaminant. On average, 90% of the 2–7 million gallons of fracturing fluid used for each well does not resurface after the injection process, and it is largely unknown where this fluid ends up. Studies of the Marcellus shale formation show that the shale is very dry, and it is likely that most fluid absorbs into the shale. There is the possibility that water that is not absorbed could move along abandoned oil and gas well shafts that are in the same region. However, study of 233 drinking wells in the shale gas region of Pennsylvania found no fluid contamination of groundwater, and even sites with dissolved methane present had no evidence of fracturing fluid. No studies included in this review had reliable evidence that contamination from fracturing fluid has happened in the past or will happen in the future.

The water that is recovered from hydraulic fracturing is a combination of injected fracturing fluid and water from underground formations. The average volume of wastewater from Marcellus wells in Pennsylvania was 26 million barrels per year between 2008 and 2011, and wastewater from Marcellus shale gas wells made up 79% of all oil and gas wastewater requiring management in

Pennsylvania in 2011. Marcellus shale wastewater is high in total dissolved solids (TDS) and radiogenic compounds. Currently almost all flowback water (recovered fracturing fluid) is captured at the surface and used in another fracturing operation after dilution or treatment. This practice of reuse greatly reduces the volume of wastewater that must be treated and disposed of, but will do so only while new wells are made at high rates. After some time, there will be more wastewater from operating wells than new well drilling will require. Reuse concentrates certain contaminants, including radioactive radium, which makes it increasingly difficult to properly dispose of reused wastewater.

Most final disposal happens in injection disposal wells, but this may not be an appropriate solution in Pennsylvania due to low injection well capacity. Currently there are only five operating injection disposal wells in Pennsylvania that are licensed to handle the level of contamination present in shale gas wastewater, and construction of new disposal wells would require regulator permission and expensive construction. Additionally, injection wells are known to induce seismic events. Until recently, municipal waste treatment facilities (also known as publicly operated treatment works or POTWs) have been used to treat wastewater, but these did not have the capacity to treat water high in TDS. Thus, use of POTWs led to an increase of TDS in waterways, which evoked significant public resistance to POTW use. Eventually PADEP created discharge limits in order to discourage the use of POTWs.

Wastewater can contaminate surface water if it is released into the environment without adequate treatment from waste treatment facilities or when wastewater is spilled from storage facilities or trucks. There are a few ways to test for surface water contamination by wastewater. Researchers look for high concentrations of sodium, calcium, and chlorine, which are present in wastewater but can originate from many sources. High concentrations of strontium, bromine, barium, and a high ratio of certain strontium isotopes are stronger evidence of wastewater contamination. Despite these indicators, research that attempts to identify the source of surface water contamination finds difficulty connecting results to Marcellus wells because there is little information available regarding pre-shale gas era water quality. To increase availability of data, PADEP started a surface water monitoring program in 2011, but currently research is underfunded and more research into proper disposal methods is necessary.

Methane Migration from Shale Gas Extraction Contaminates Drinking Water in Pennsylvania

Perhaps the biggest environmental and health concern related to shale gas development is the possibility of contaminants leaking from the well shaft into nearby groundwater supplies. The first sign of such leakage would be stray methane in groundwater, as methane is a small enough molecule to move through tiny spaces and easily dissolves in water. Jackson *et al.* explored the possibility of stray gas contamination by testing for concentrations of methane, ethane, and propane in drinking water wells of homes in the Marcellus shale region of Pennsylvania. The researchers generally found higher amounts of dissolved gases in drinking water wells less than one kilometer from a natural gas well. Statistical analysis showed that distance from gas wells was the most significant factor for raised levels of natural gas. Other possible sources of natural gas contamination—valley bottom streams and the Appalachian Structural Front—were ruled out using multiple regression analysis, linear regression, and Pearson and Spearman coefficients. In addition, the authors tested the chemical composition of the gas samples to see if their source was biogenic, i.e. produced by microorganisms, or thermogenic, i.e. with a potential connection to shale gas production. This chemical analysis showed that at least some of the natural gases present in drinking water wells came from a thermogenic source. The authors suggest that any stray gases could be due to wells with faulty steel casings or cement seals.

Jackson *et al.* sampled 81 drinking water wells and combined their results with information from 60 previously-collected samples. The study found dissolved methane at 115 of 141 homes (82%), ethane at 40 of 133 homes (30%), and propane at 10 of 133 homes (8%). Methane concentrations were high in all cases relative to ethane and propane concentrations, and homes within one kilometer of a natural gas well had 6 times the amount of methane as homes farther away. The 12 highest concentrations of methane were above the level at which the US Department of Interior recommends hazard mitigation, and 11 of those houses were within one kilometer of a gas well. In addition, homes within one kilometer of a gas well had 23 times the amount of ethane as homes farther away, and propane was only detected at homes within one kilometer of a gas well. Ethane and propane only derive from thermogenic sources, so their presence is evidence that the natural gas contamination is likely from a gas well.

Another way to determine if gas is from a thermogenic source is to look at its isotopic signature. Yet again the strongest evidence for thermogenic sources (the most $\delta^{13}C–CH_4$ signatures greater than –40‰) were within one kilometer from natural gas wells. Shale gas has a specific trend in which heavy carbon isotopes in methane ($\delta^{13}C$) are more abundant than those in ethane, whereas the reverse is usually true. Six out of 11 houses where sampling was possible showed this trend, indicating that those gas samples came from shale gas extraction sites.

The helium isotope ^4He is a component of thermogenic natural gas. The ratio of this isotope to methane (^4He to CH_4) in the dataset was fairly consistent, except for the points with elevated levels of methane. These had a ratio of ^4He to CH_4 that was consistent with Marcellus production gases, which is somewhat lower than normal drinking water levels.

The researchers believe that poor well construction could have led to this contamination of drinking water. In particular, stray gases could have escaped through faulty protective steel casings or from imperfections in the cement sealing between the casings and rock outside of the well. Faulty steel casings would lead to gas from inside the well leaking out to the surroundings, followed by metals, fracturing fluids, or other evidence of gas extraction. Faulty cement would lead any gas in the spaces around the well to escape upwards into drinking water, meaning the gases would not be easily identifiable with a particular well, as was the case with the samples recorded in this study.

The authors would like to see further research go towards understanding more about how drinking water quality near the Marcellus shale gas production area compares to drinking water quality near other shale gas sources. They also suggest gathering pre-drilling data in order to make detailed studies of water quality before, during, and after drilling and hydraulic fracturing.

Trends in Wastewater Management Practices since 2008 in the Marcellus Shale Region

Since 2008, the Marcellus shale formation has become the most productive region for extracting shale gas in the US. Managing wastewater for these operations is a challenge not only due to their size and distribution, but also because of the different types of contaminants that are present in various types of wastewater. Rahm *et al.* (2013) retrieved data from the Pennsylvania Department of Environmental Protection (PADEP) Oil and Gas Reporting website from 2008 to 2011 to look for the trends and drivers of Marcellus shale wastewater management. After analysis using internet resources and Geographic Information Systems (GIS), the authors found that there was a statewide shift towards wastewater reuse and injection disposal treatment methods and away from publicly owned treatment works (POTW) use. These wastewater management trends are likely due to new regulations and policies, media and public scrutiny, and natural gas prices. Research also shows that Marcellus shale development has influenced conventional gas wastewater practices and led to more efficient wastewater transportation.

Rahm *et al.* retrieved data from the publicly accessible PADEP Oil and Gas Reporting Website for the years 2008 through 2011. Information was submitted to this website by industry operators, so there is definitely potential for human error. Wastewater disposal methods reported to the website included industrial waste treatment plants, municipal sewage treatment plants, injection disposal wells, reuse (which involves some industrial treatment), and other or not determined. The authors also gathered information about conventional wastewater management for 2004–2006 and 2010–2011, which approximately represent the pre- and post- Marcellus shale development years. Information from the database was used to look up and geographically locate disposal facilities in order to calculate wastewater transportation miles. The shortest driving distance from wastewater generation site to wastewater destination was calculated using Google Earth. The authors made different analyses for statewide water disposal in Pennsylvania and for two regions of the Marcellus Shale.

The most significant trend was the increased reuse of wastewater and the decreased utilization of POTWs. New regulations and public pressure in large part drove this change. In 2009, PADEP required the submittal of a formal discharge strategy for wastewater high in total dissolved solids (TDS). This requirement limited the options for wastewater disposal, as POTWs usually cannot treat high-TDS waste. Public concern towards shale gas development and opposition to the use of public facilities for wastewater led to the state of Pennsylvania taking further action. Act 15, passed by the state legislature in 2010, made rules for tracking residual waste, required operators submit a report on gas production and wastewater management every six months, and established a database so the public could access this information. In response to continuing public pressure, PADEP requested in 2011 that operators completely cease the use of POTWs for wastewater management.

Natural gas prices influenced the trend of increased of injection well usage. When gas prices decreased between late 2011 and early 2012, fewer new wells were drilled and there was less of a demand for recycled water. Though reuse was by far the largest "disposal" method for wastewater, injection well usage increased substantially throughout the test period.

Conventional gas wastewater had additional disposal options and a relatively larger use of POTWs before major Marcellus shale development. After 2010, however, wastewater disposal methods for conventional and unconventional gas came to resemble each other. This is likely due to public scrutiny of Marcellus development, which pressured regulators to monitor disposal and limit treatment options for both conventional and unconventional gas producers. New regulations after 2010 affected conventional gas wastewater reporting, treatment, and discharge, and called for information related to wastewater to be publicly accessible. Another driver for the shift in conventional gas wastewater management practices was the infrastructural development in response to the Marcellus boom that led to increased industrial treatment capacity for all natural gas producers.

The final trends were related to wastewater transport. Marcellus shale infrastructure is currently quite spread out, evidenced by wastewater shipments to various areas throughout Pennsylvania, Ohio, West Virginia and Maryland. Infrastructure tends to develop regionally so that large wastewater exporters are near wastewater importers. An example of this is the southwest region of Pennsylvania, which had a high rate of injection disposal relative to other Marcellus drilling regions due to its proximity to the wastewater importing areas of Ohio and West Virginia. The presence of nearby infrastructure led to faster shale gas development in the southwest region, even in times of low natural gas prices. The northeast region did not have as much infrastructure before the Marcellus boom, so their wastewater travelled farther at first. As infrastructure developed across the state, wastewater travelled on average 30% less distance in 2011 than in 2008.

Wastewater Injection Sites Shown to be Susceptible to Remote Earthquake Triggering

Underground fluid injection can induce earthquakes by increasing pressure on the geologic features deep below the surface of the earth. A growing number of sites in the past decade have shown signs of induced seismicity due to an increase in subsurface wastewater disposal associated with natural gas extraction. Induced seismic events may not appear for a few days or up to many years, and there is currently no way to diagnose underground damage near fluid injection sites in order to predict such events. Van der Elst *et al.* (2013) investigated the possibility that remote earthquake triggering from large, distant earthquakes may be an indicator of stress for wastewater injection sites. The triggers used for this study were three earthquakes with strengths larger than 8 moment magnitude (M_w) between 2010 and 2012. The authors found three sites with a history of fluid injection that responded to these triggers with patterns of seismicity indicative of underground damage.

Van der Elst *et al.* (2013) examined earthquake data to look for sites that responded to three large trigger earthquakes. The earthquakes were a February 2010 8.8 M_w earthquake in in Maule, Chile; a March 2011 9.1 M_w earthquake in Tohoku-oki; and April 2012 8.6 M_w earthquake in Sumatra. In order to identify sites where seismicity was induced, researchers collected data for all earthquakes 3 M_w and higher in the central United States that occurred within 10 days of each event. When the authors mapped all 3 M_w and higher earthquakes that occurred within 10 days of a trigger earthquake, they found that triggering happened almost exclusively at three sites: Prague, Oklahoma; Snyder, Texas; and Trinidad, Colorado. Additional research showed that each of these sites had relatively low seismicity before the first trigger, at least one fairly large earthquake in response to the trigger, and a delayed earthquake "swarm" months after the trigger. Additionally, each site had a strong history of wastewater injection within 10 kilometers of the induced earthquake activity.

The first site was the Cogdell oil field near Snyder, Texas. This site had a number of earthquakes in response to the March 2011 event in Tohoku-oki. The largest response was a *3.8* M_w event that happened two and a half days after smaller events. A few months later, in September 2011, the site had a seismic swarm that included a 4.3 M_w main shock. The rate of earthquakes was higher at the site in the 10 days after the earthquake at Tohoku-oki and immediately after the September swarm than at other any time from February 2009 to the present.

The 2010 Maule event triggered a series of earthquakes near a fluid injection site in Prague, Oklahoma. The largest event was 4.6 M_w, and it occurred only 16 hours after the Maule earthquake. There was a very low rate of earthquakes in the site's history prior to the trigger and no activity measured in the 4 months before the Maule event. These triggered earthquakes were suggestively located near the epicenter of a 5.0 M_w earthquake that occurred the following year, in November 2011. This event led to two more earthquakes with a magnitude greater than 5.0 M_w. The largest of these, a 5.7 M_w in November, could possibly be the largest earthquake associated with wastewater disposal ever recorded. As a sign of continued seismicity, the 2012 earthquake in Sumatra caused a small amount of activity near the edge of the swarm that occurred in 2011.

Trinidad, Colorado also experienced seismic activity as a result of the Maule earthquake. There were only four events in the day after the earthquake in Maule, but the site only had five earthquakes in the entire previous year, so this result is significant. The site's delayed swarm occurred in August 2011 and included a 5.3 M_w main shock. As in Oklahoma, the Sumatra earthquake caused a small boost in earthquake activity in Trinidad near the site of the previous earthquake swarm.

The authors are concerned that these findings indicate the likelihood of future induced seismic activity at these damaged sites, and they suggest that improved seismic monitoring should occur in areas of subsurface wastewater injection.

Shale Gas Well Drilling and Wastewater Treatment Impacts on Surface Water Quality in Pennsylvania

Shale gas development can affect surface water quality by means of runoff from well construction and discharge from wastewater treatment facilities. Olmstead *et al.* (2013) conducted a large-scale statistical study of the extent to which these two activities affect surface water quality downstream. This study is different than most current literature related to the regional water impacts of shale gas development in that it focuses on impacts to surface water bodies as opposed to groundwater bodies. Researchers consulted online databases to retrieve locations of shale gas wells and wastewater treatment facilities within Pennsylvania. These were spatially related to downstream water quality monitors using Geographic Information Systems (GIS). Concentrations of chloride (Cl^-) and total suspended solids (TSS) were used as indicators of water quality because both are associated with shale gas development and are measured by water quality monitors. Shale gas wastewater typically has a high concentration of Cl^-, which can directly damage aquatic ecosystems and is not easily removed once dissolved in water. TSS, which harm water quality by increasing temperature and reducing clarity, can potentially come from the construction of well pads, pipelines, and roads associated with well drilling, especially when precipitation creates sediment runoff. Results of the study suggest that wastewater treatment facilities are responsible for raised concentrations of Cl^- downstream and that the presence of gas wells are correlated with raised concentrations of TSS downstream.

Olmstead, *et al.* collected publicly-available data from national and Pennsylvania state websites. TSS and Cl^- concentration observations came from the Storage and Retrieval Data Warehouse (STORET) database of the US EPA. The locations and permitting and drilling dates of shale gas wells came from the websites of the Pennsylvania Department of Environmental Protection (PADEP) and the Pennsylvania Department of Conservation and Natural Resources (PADCNR). Shipments of wastewater and the locations of their destination treatment plants also came from PADEP. In order to test for the impacts of treated wastewater on surface water Cl^- concentrations, the authors measured the density of wastewater treatment facilities and the volume of total wastewater shipments upstream of water quality monitors. Wastewater shipments to both publicly owned treatment works (POTWs) and centralized waste treatment (CWT) facilities were included, as long as they had NPDES permits making them eligible to treat shale gas waste. An additional cause of Cl^- contamination is accidental release of wastewater during the hydraulic fracturing process, so density of upstream shale gas wells was also considered. TSS contamination models tested for impacts from waste disposal and from well construction, focusing on the time period when land clearing and well pad construction take place. Controlled factors included precipitation and seasonal changes in contaminant levels that are considered "normal" in different watersheds. For example, road salt is a source of Cl^- in certain times of year in certain watersheds. The authors controlled for these "fixed effects," thus eliminating other potential sources of raised contamination levels besides shale gas development.

The most significant statistical factor on raised downstream Cl^- concentrations is the density of upstream wastewater treatment facilities. An increase of 1.5 facilities per watershed (one standard deviation) increases Cl^- concentrations by 10–11%. In general, there was a weak positive correlation between raised levels of Cl^- and the quantity of wastewater treated upstream. This result reflects the fact that many wastewater treatment plants do not have the capacity to treat very high concentrations of Cl^-. The density of upstream shale gas wells have an insignificant effect on Cl^- concentration, even when measurements are limited to wells that were developed within 0–3 and 3–6 months before a water sample was taken. This limitation should account for accidental releases of hydraulic fracturing fluid.

The presence of shale gas wells in a monitor's watershed correlated with high concentrations of TSS, while the presence of waste treatment facilities had a statistically insignificant effect on downstream TSS. Increasing the average density of well pads by one standard deviation—or 18 well pads per watershed—leads to a 5% increase in TSS concentrations in surface water. The specific relationship between well pad density and downstream TSS concentration remains unclear. It seems likely that well pad construction would be a cause of raised TSS concentrations, however the data showed no change in downstream concentrations when water quality measurements are limited to the time window when well pads are in construction. TSS concentrations also do not increase when well pad density is considered along with precipitation, even though a change would be expected if TSS are transported from construction sites by stormwater. Further research could determine if road or pipeline construction, spills, or other emissions from well sites are responsible for the measured increase in TSS. Further research is also needed to investigate costs, benefits, and goals for controlling water contamination levels due to shale gas development impacts.

Challenges of Regulating the Disclosure of Hydraulic Fracturing Chemicals

Hydraulic fracturing requires a large quantity of fluid; most estimates place the amount at 2 to 4 million gallons per well. This fluid is composed of 90% water, 8–9.5% proppants (sand which is needed to keep fractures open once hydraulic fracturing occurs), and 0.5–2% chemicals. Companies that perform hydraulic fracturing invest time and resources into creating their fracturing fluid formulas, so they insist on keeping those formulas proprietary because revealing the information could cause the company to lose its competitive edge. However, some common fracturing chemicals have been identified and are known to cause adverse human health effects, so keeping the composition of fracturing fluid confidential could be dangerous in the case of an emergency situation. Even in normal operating circumstances, these fracturing fluids could theoretically make their way into surface water or groundwater because hydraulic fracturing creates new flow paths through deep shale formations and speeds up the natural flow of fluids closer to the surface or aquifers. Maule *et al.* (2013) investigated recent efforts to regulate the disclosure of fracturing chemicals. Current systems in place include a voluntary reporting website, limited state regulation, and no federal regulation. Regulation efforts have faced problems of exemptions or loopholes, inadequate or incomplete information reporting, lack of enforcement, and competing state and federal interests.

Maule *et al.* investigated and analyzed voluntary fracturing fluid chemical disclosure efforts, federal regulation attempts, and current regulations in Texas and Pennsylvania. Efforts to encourage voluntary reporting of fracturing fluid composition began when the Groundwater Protection Council and Interstate Oil and Gas Compact created FracFocus.org in 2011 as a website for industry operators to self-report well-specific fracturing fluid information. Industry and some state governments view this system as adequate. However this system has a number of problems, the primary one being that it is completely voluntary so there is no way to guarantee complete or accurate information. It is common for industry operators to report only the general class of chemical used instead of the exact chemical identity, and there is no government oversight to guarantee that information is sufficient for lawmakers, regulators, and communities.

Current federal programs meant to protect the public against contamination and misinformation have loopholes and exemptions that allow hydraulic fracturing to go largely unregulated. One case is the Emergency Planning and Community Right-to-Know Act (EPCRA). This act created the Toxic Release Inventory (TRI), which was intended to make information about intentional and unintentional discharge of toxic chemicals available to the public. Hydraulic fracturing is exempt from reporting to this inventory because "Oil and Gas Extraction" is not included in the list of industries that need to report to the TRI. Another program that could be interpreted as responsible for hydraulic fracturing is the Underground Injection Control (UIC) Program of the Safe Drinking Water Act (SDWA), which requires injection wells to undergo permitting, reporting, and monitoring. The US EPA does not include hydraulic fracturing under this program because they consider the whole process as an "extraction" instead of an "injection" process. In 1997, the Legal Environmental Assistance Fund sued the EPA in order to reclassify wells that perform hydraulic fracturing as a particular type of injection well. The court's decision required the EPA to do a study to assess the risk posed to human health by hydraulic fracturing. In 2004, the EPA published a report stating that no further study was necessary. However, this report utilized only existing literature (which was extremely limited) and interviews with industry and government representatives. Additionally, it only considered the effects of drilling in coal beds, while hydraulic fracturing happens within a variety of geological formations. This highlights the recurring pattern in hydraulic fracturing regulation in which lack of information leads to lack of investigation and regulation; lack of investigation and regulation leads to lack of information. Because no conclusions could be drawn from the 2004 EPA study, an Energy Policy Act that Congress passed in 2005 exempted hydraulic fracturing from regulation under the Safe Drinking Water Act.

Two acts introduced in Congress in the last five years have explicitly attempted to put regulation of hydraulic fracturing in federal hands, but neither was passed into law. The American Power Act, introduced in 2010 by Senators John Kerry and Joseph Liebermen, would have amended EPCRA to include hydraulic fracturing. The bill required hydraulic fracturing companies to disclose to the public all chemicals used in each fracturing operation. For reasons unrelated to the hydraulic fracturing amendment clause, the Act was opposed and did not make it out of the committee level. The Fracturing Responsibility and Awareness (FRAC) Act was under consideration in House and Senate committees during the 111th and 112th Congressional Sessions. Had it passed, this act would have removed the clause of the SDWA that exempts hydraulic fracturing from the UIC program. Hydraulic fracturing operators would have been required to disclose to a designated government regulator the identity and volumes of the intended chemicals before injection and the actual chemicals injected into the well. Information about nonproprietary chemicals would have been released to the public, but proprietary chemicals would not have faced public disclosure under any circumstances. In the case of an emergency, regulators and emergency responders would be informed of the identity of all chemicals, but the public would not. Therefore, FRAC would have made more information available to the public but kept trade secrets hidden, and protected the public in case of emergency. FRAC was very contentious; industry argued that it would force them to reveal too much proprietary information, state governments didn't want federal agents to have control, and environmental groups pushed for even more stringent regulation. The bill never made it out of committee.

State regulations in Texas and Pennsylvania at present require companies to disclose the chemical compositions of fracturing fluids, but have limited control and significant loopholes. Texas, which has a long history of fighting federal control, enacted fracturing disclosure regulation in 2012 in order to preempt the FRAC Act while it was in Congress. These regulations require disclosure of nonproprietary chemicals in fracturing fluids for wells with permits issued after February 2012. Disclosure is only required after completion of a fracturing treatment, not before, and unintended releases do not need to be reported. Proprietary chemicals need only be reported in an emergency situation. Also, it is only required to report the maximum concentration of each chemical, not its total volume. The information is put on the internet by industry operators and the Railroad Commission (which regulates much of Texas' oil and gas industry) is responsible for enforcing disclosure. However, industry operators are not held responsible for reporting inaccurate information, and the lack of enforcement capability allows information to go missing or unreported. In fact, it is estimated that half of new wells drilled in Texas do not have disclosure reports submitted to the online registry. The Pennsylvania Legislature followed Texas' lead by updating their 1984 Oil and Gas Act in early 2012 to include a requirement that well operators submit a chemical disclosure form and post to a registry within 60 days after drilling activities. The updated regulations have the similar loopholes to their Texas equivalents, and incomplete or incomplete information reporting is common. Enforcement of disclosure is extremely difficult due to the fact that there are simply not enough regulators to handle the thousands of new well operations that occur each year.

Much of the difficulty with regulation stems from the clash between state and federal authorities. States argue that fracturing operations only affect in-state interests, so states should regulate them without federal intervention. A new act in Congress, Fracturing Regulations are Effective in State Hands (FRESH) Act, acknowledges this perspective. If passed, this act would guarantee that states would have exclusive authority to regulate the fracturing operations within state boundaries. However, Maule *et al.* argue that federal oversight may be necessary in order to ensure sufficient and accurate reporting.

Some government officials and environmental groups also recognize that federal involvement may be necessary in this matter, but all efforts have been progressing slowly. In 2011, the Shale Gas Production Subcommittee of the Secretary of Energy Advisory Board (SEAB) recommended that disclosure of fracturing fluid composition be implemented immediately. They stated that the voluntary reporting on FracFocus.org must be advanced to include reporting of proprietary chemicals and have well-specific information available to the public. Though forward-thinking, the SEAB recommendation did not specify a plan to implement disclosure and instead delegated that responsibility to the Department of Interior (DOI). Other current efforts attempt to include hydraulic fracturing under major environmental laws. For example, in 2011 the environmental group Earthjustice petitioned the DOI to include hydraulic fracturing under the Toxic Substances Control Act (TSCA) Section 4, which states that the EPA can require testing of chemicals which may either present significant human health risk or which are produced in high quantities. The EPA responded by saying there is not explicit evidence that hydraulic fracturing chemicals fall under either of TSCA Section 4's requirements. Of course, the lack of evidence is due to the lack of disclosure and research, which were the cause of slow regulatory progress throughout this investigation.

Shale Gas Development Poses Threats to Regional Biodiversity

Shale gas development physically and chemically alters the surrounding landscape, and native plants and animals can be particularly susceptible to these changes. In the Marcellus and Utica shale region—a largely forested area that encompasses the states of Pennsylvania, Ohio, and West Virginia—shale gas wells are being drilled with increasing density. A shale gas installation, including the well pad, compressor station, and storage areas, requires 3.56 ha on average. If an edge effect is considered, installations can affect approximately 15 ha of forest per site. Kiviat (2013) reviewed the potential ways that shale gas development can impact biodiversity. The most serious threats are physical alteration of terrestrial landscapes, chemical contamination of water and soil, and alteration of regional hydrology. Terrestrial alterations include construction of well installations, which cause deforestation and habitat loss, and construction of roads and pipelines, which create forest fragmentation. Chemical contaminants come from fracturing fluid and recovered wastewater. Hydrologic alterations are caused by water withdrawals and an increase in impermeable surfaces. Minor impacts on plant and animal health can come from noise, light, and air quality. Certain species are particularly at risk from shale gas development activities and some are able to thrive in the altered conditions.

Kiviat conducted a literature review to analyze the consequences of various shale gas development activities on nearby plants and animals, and identified species that might be impacted positively or negatively from shale gas development. The first major consequence of shale gas drilling is physical alteration of the land from the construction of well pads, roads, and pipelines. The construction of a well pad destroys any organisms that had occupied the space prior to construction and contributes to habitat loss and deforestation. Edge effects from well pads range from 10 meters for trees to as much as 500 meters for birds. These installations, along with their associated roads and pipelines, fragment the surrounding forested area. Fragmentation affects seed dispersal, pollination, predator-prey relationships, and herbivore-plant relationships. These factors are crucial to the functioning of local ecosystems, and if disturbed will particularly harm orchids, herbs, lichens, amphibians, birds, and a certain species of butterfly. Roads and vehicular traffic additionally create the possibility of road mortality, especially with frogs. Certain species of nonnative plants, on the other hand, can thrive in these fragmented and industrial conditions. Many weeds disperse along roads; others will grow in the disturbed soils at the edges of roads, pipelines, or well pads and will spread from there into the nearby forests. Other effects from well pad construction are the warming and drying of remaining forest, which create favorable conditions for certain nonnative plants and songbird nest predators. The new conditions may also make a landscape that is favorable for deer, which threaten herb populations. It will be difficult for these forest ecosystems to recover even after wells come to the end of their useful life (usually around 20-40 years) and the surrounding area is remediated, as it can take 75–100 years or more for forests to regenerate and mature.

The hydraulic fracturing process can negatively impact biodiversity by adding toxic chemicals to regional water and soil, which happens when accidental spills or leaks of fracturing fluid or wastewater occur. Such spills release VOCs, diesel, metals, sodium chloride, and many other substances to nearby surface water. Sodium chloride is especially a problem for biodiversity as many streams are already high in salt content and many amphibians, lichens, mosses, conifers, and aquatic plants are sensitive to salt. When hydraulic fracturing fluid resurfaces as toxic and radioactive wastewater, it is stored in open-air ponds that are potential ecological traps for water birds, muskrat, turtles, frogs, and aquatic insects. Water quality can also decline due to sediment pollution from heavy equipment on rural roads or by inadequate erosion control at drill sites. In one study in Arkansas, stream turbidity increased with shale well density.

Hydrologic alteration is another side effect of shale gas development. Significant water withdrawals for hydraulic fracturing fluid may harm stream fishes and aquatic invertebrates that require a minimum level of water throughout the year. Replacing forest with impermeable well pads increases total storm water runoff, which can decrease water quality and species diversity in streams. Freshwater mussels and aquatic salamanders are two species that are known to be particularly sensitive to hydrological conditions and could be negatively impacted by these changes. Groundwater tables and their flow into streams and wetlands might also be affected by this changing hydrology.

Minor impacts come from noise, light, and air emissions. Diesel compressors on shale gas well sites create noise 24 hours a day. Loud noises are known to interfere with acoustic communication of frogs, birds, and mammals. They can also cause hearing loss, stress, and hypertension in many animals, and bats tend to avoid loud noises altogether. These effects can lead to an overall change in population composition if mating success is disturbed. Well pads are lit throughout the night and sometimes have continuous artificial lighting. These lights can attract and kill moths and aquatic insects. Other animals can have their mortality, reproduction, and foraging affected positively or negatively by artificial light. Diesel exhaust, VOCs in fracturing fluids, ground level ozone, and road dust can negatively affect air quality. This will likely harm nearby moss and lichen population, but not much research has been done to see how it can affect animals.

Range-restricted species are particularly vulnerable to shale gas development. One study showed that 15 plants and animals' geographic ranges overlap the Marcellus and Utica shale region by at least 36%. The study notes that some species whose ranges have not been mapped may be quasi-endemic or have seasonal habitats in this region. Alternatively, some native organisms may thrive in the new habitats created by shale gas development. Bare and disturbed soils can be nest sites for bees, wasps, reptiles, and birds. Snakes are attracted to warm pavement in cold weather; they and other animal species will likely thrive in the warmer microclimate around well pads. Metal-tolerant plants and mosses could also be very successful in these new industrial climates.

Shale gas development is not the only risk to biodiversity in the Marcellus and Utica shale region; there can be similar impacts from coal mining, logging, urban sprawl, agriculture, and climate change. These factors will probably act synergistically with shale gas development and cause a more serious problem than current research indicates. Shale gas development is a unique problem in its use of toxic chemicals, rapid development, and geographic extent. The authors recommend careful management of chemicals, wastewater, soil, and other pollutants in order to protect regional biodiversity. One technique that can be utilized in order to minimize risks to biodiversity is reusing wastewater, which reduces total withdrawals from local water resources and reduces vehicle traffic necessary to transport water in and out of a site. Another technique is drilling further horizontally underground to reduce the total number of well pads and limit fragmentation. Placing well pads closer to highways or on land already lacking in biodiversity would also be beneficial. Finally, the researchers note that in order to remediate efficiently, post-well activities should be considered on a landscape level, not site by site.

According to Life Cycle Assessment, Shale Gas Produces Half the GHG Emissions and Consumes Half the Freshwater of Coal

The long term environmental concerns having to do with shale gas development are primarily greenhouse gas (GHG) emissions and freshwater consumption, as other forms of environmental degradation can be remediated over time. Ian Laurenzi and Jersey Gilbert (2013) of the ExxonMobile Research and Engineering Company performed a life cycle assessment (LCA) of both GHG emissions and freshwater consumption of Marcellus shale gas. The life cycle begins with well drilling and ends with burning the fuel for power generation. Using their elaborated system boundaries, the researchers found that a Marcellus shale gas well releases 466 kg of carbon equivalent units per megawatt hour of power produced (kg CO_2eq/MWh) and consumes 224 gallons of freshwater per megawatt hour of power produced (gal/MWh) over the course of its lifetime. The biggest contributor to both GHG emissions and freshwater consumption is power generation. The result of this study are highly dependent on the variables chosen to represent the shale gas well life cycle, especially the expected ultimate recovery of natural gas. Despite the potential for variability of results, the result of 466 CO_2eq/MWh is consistent with other published life cycle assessments for conventional and shale gas, and almost all of the 14 other studies fall within the 10%–90% range of 450–567 CO_2eq/MWh. Even considering factors that can increase total results, this study shows that average GHG emissions from shale gas are 53% lower and freshwater consumption is 50% lower than what is required for an average coal life cycle.

Laurenzi and Jersey used a "from well to wire" approach to study the carbon and water footprints of Marcellus gas. In this study, the shale gas life cycle was defined to include drilling, well completion, wastewater disposal, transportation of gas from well via gathering pipelines, treatment and processing, transmission, and power generation. It considered water consumed for hydraulic fracturing and evaporative cooling at the power plant. Excluded from the study were gas distribution networks, which deliver gas for purposes other than electricity. The GHG emissions are expressed in units of CO_2 equivalents based on an IPCC specification. The purpose for this unit is to make the "global warming potential" for all greenhouse gases comparable. This study used 100-year global warming potential values of 25 kg CO_2eq/kg for methane (CH_4) and 298 kg CO_2eq/kg for nitrous oxide (N_2O). The functional units for the whole study were kg CO_2eq/MWh (amount of gas released per unit of power produced) and gal/MWh (gallons of water consumed per unit of power produced). These units mean that the study did not report GHG emissions and power consumption as a gross total amount but as a ratio of resources inputted to power outputted. The authors used data from over 200 Marcellus shale wells in West Virginia and Pennsylvania. Their information largely came from XTO Energy, a subsidiary of ExxonMobil, and where data were not available they used established standards from different regulatory agencies or publicly available data. Modeling of the power generation stage used a combined cycle gas turbine (CCGT) power plant operating at 50.2% efficiency.

The researchers' calculations revealed that the total life cycle GHG emissions of a Marcellus shale gas well are 466 kg CO_2eq/MWh. Almost 78% of emissions occur at the power plant, where the shale gas is burned to create electricity. The second most significant source of GHG emissions are the gas engines that drive the gathering system compressors, which are part of the network that transports gas from various wells to a central location. Hydraulic fracturing activities are only responsible for 1.17% of the life cycle GHG emissions. As hydraulic fracturing is the major difference between shale gas and conventional gas life cycles, only 1.17% of GHG emissions are specific to Marcellus shale gas production and processing, making the difference between Marcellus shale gas and conventional gas emissions statistically insignificant. Some other sources of emissions are: transmission compressors, transmission losses, processing plant compressors, processing losses, pneumatic devices and chemical injection pumps, and road transportation for well maintenance.

The total life cycle water consumption is 224 gal/MWh, with 93.3% of that total occurring at the power plant, where water is used for cooling. Hydraulic fracturing requires 6.2% of the total (13.7 gal/MWh), representing the majority of the water consumed before the power plant. That figure takes into account the water used in the life cycles of gasoline or diesel that are used to power the fracturing process or vehicular transportation. The final 0.5% of water is consumed during drilling, casing manufacture, and road transportation for well maintenance.

Researchers ran a simulation that varied different parameters of the LCA for GHG emissions and measured how much the final results changed.

This simulation determined that life cycle GHG emissions are most dependent on the expected ultimate recovery of natural gas from a well. This reflects the fact that the initial investment of resources for well drilling and completion will yield more power over the course of the well's lifetime if more natural gas is recovered from the well. Other important parameters are pipeline length, gas composition, water scarcity in the region, and other infrastructural elements.

The efficiency of the power plant is a factor that could greatly change the final result, as it is dependent on power output. This assessment assumed an efficiency of 50.2%, which is relatively consistent with the 80% of US power plants that operate within the range of 42–48% efficiency. An additional life cycle assessment using the efficiencies of the currently operating US CCGT fleet resulted in a distribution that was wider with a higher amount of GHG emissions. Even so, comparing the results of this higher, more realistic life cycle distribution to the GHG emissions of an average coal life cycle (calculated by a different research group) shows that the carbon footprint of Marcellus shale gas is approximately 53% that of coal. The highest possible level of Marcellus shale gas emissions from this higher life cycle distribution is lower than the lowest possible GHG emission level of coal. Using the same LCA for coal, researchers determined that the freshwater consumption of shale gas is approximately 50% that of coal.

Conclusions

This review has summarized current research of the major environmental implications of shale gas development, focusing on issues related to regional water quality and wastewater disposal. These studies have shown that shale gas development poses a number of risks to the surrounding environment, however no studies prove significant damage directly caused by shale gas development. This is a difficult connection to make because pre-drilling information is not available, monitoring is not currently required, publicly-available data is reported on a voluntary basis, and chemical compounds involved with hydraulic fracturing remain legally confidential. Further research is needed to more accurately determine the exact effects of shale gas development on the surrounding environment, though many believe this will be unlikely without government intervention to regulate industry practices.

References Cited

Jackson, R., Vengosh, A., Darrah, T., Warner, N., Down, A., Poreda, R., Osborn, S., Zhao, K., Karr, J., 2013. Increased stray gas abundance in a subset of drinking water wells near Marcellus shale gas extraction. Proceedings of the National Academy of Sciences 110, 11250–11255.

Kiviat, E., 2013. Risks to biodiversity from hydraulic fracturing for natural gas in the Marcellus and Utica shales. Annals of the New York Academy of Sciences 1286, 1–14.

Laurenzi, I., Jersey, G., 2013. Life cycle greenhouse gas emissions and freshwater consumption of Marcellus shale gas. Environmental Science and Technology 47, 4896–4903.

Maule, A., Makey, C., Benson, E., Burrows, I., Scammell, M., 2013. Disclosure of hydraulic fracturing fluid chemical additives: Analysis of regulations. New Solutions 23, 167–187.

Olmstead, S., Muehlenbachs, L., Shih, J., Chu, Z., Krupnick, A., 2013. Shale gas development impacts on surface water quality in Pennsylvania. Proceedings of the National Academy of Sciences 110, 4962–4967.

Rahm, B., Bates, J., Bertoia, L., Galford, A. Yoxtheimer, D., Riha, S., 2013. Wastewater management and Marcellus shale gas development: Trends, drivers, and planning implications. Journal of Environmental Management 120, 105–113.

Van der Elst, N., Savage, H., Keranen, K., Abers, G., 2013. Enhanced remote earthquake triggering at fluid-injection sites in the midwestern United States. Science 341, 164–167.

Vidic, R., Brantley, S., Vandenbossche, J., Yoxtheimer, D., Abad, J., 2013. Impact of shale gas development on regional water quality. Science 340, 777–892

2. Sustainable Bioenergy in the Age of Climate Change

Christina Whalen

Growing social, economic, and environmental concerns about the world's unsustainable petroleum consumption has led to the need for quick and effective conversion to biofuel sources. Biofuels are beginning to enter the energy market as the need for energy security increases, oil prices continue to increase, and our natural resources continue to deplete. The most common first generation biofuels, ethanol and biodiesel, are derived from conventional food crops like starch, sugar, and oil from crops, such as wheat, maize, sugar cane, palm oil, and oilseed rape. Sustainable biofuels will help increase market competition and moderate oil prices, while also providing a valuable alternative energy source that will reduce our dependency on fossil fuels and aid in mitigating the adverse effects of climate change and global warming. In the world of agriculture, there is much unused and abandoned land that could potentially be used specifically for bioenergy crop cultivation, originating from conventional crops.

An important and renowned biofuel, sugarcane, has been used to produce ethanol since the 1970s resulting from the 1973 oil crisis, when members of Organization of Arab Petroleum Exporting Countries (OAPEC) announced an oil embargo. Brazil led the world in producing the first sustainable biofuel economy. Since then it has contributed to a 61% reduction in GHG emissions, resulting in the U.S. Environmental Protection Agency (USEPA) designating it an advanced biofuel. The Brazilian sugarcane ethanol fuel program has experienced great success because of the efficient cultivation technology and modern equipment. Studies have found that sugarcane ethanol production did not raise conventional food prices, nor had significant adverse environmental effects, thus Brazilian ethanol continues to lead the biofuel industry.

Advanced biofuels, also known as second-generation biofuels, can be produced from different forms of biomass, in fact to any organic carbon source that is refurbished during the carbon cycle. They are derived from plant materials, but can also stem from animals. These plant-based biofuels, which are not yet very commercial, are derived from lignocellulosic biomass, which cannot be fermented without pre-processing, thus it is harder to extract usuable fuel from them than from first generation biofuel stock. The most common second generation biofuel stock include wheat straw, *Miscanthus*, poplar, and willow, all of approximately equal effectiveness. Other viable feedstocks include municipal solid waste, green waste, and black liquor. First generation biofuel technologies have significant limitations because not enough can be produced without threatening valuable food supplies and biodiversity. They also often depend on government subsidies, are not cost competitive with fossil fuels such as oil, and only mitigate greenhouse gas (GHG) emissions by a miniscule amount, if at all. This is why second generation biofuels have emerged as a viable substitute. These biofuels have the potential to supply a larger proportion of the nation's fuel providing sustainability, affordability, and environmental benefits, potentially reducing GHG emissions by about 90%. The technologies used in producing second-generation biofuels are gasification, pyrolysis, torrefaction, and biochemical routes. Recent commercial developments are occurring in Canada, Sweden, the United Kingdom, and Finland (2013).

Miscanthus has attracted attention as an energy source crop because of its high carbon fixation rate, low mineral content, and high biomass yield. Studies have shown *Miscanthus*' potential for producing bioethanol and thermochemical valorizations, but technological and environmental factors also need to be considered when discussing future industrial developments (Brosse *et al.)*. Similarly, the abundance of maize stover in the United States makes it a feasible source of feedstock for producing biofuels, especially liquid fuels. One of the most promising aspects of using maize is the potential to commercialize the stover as a biomass source leaving the grain as a food source. In order to increase biomass production various aspects need to be managed: plant architecture, density tolerance, resistance to antibiotic stresses, and natural acquisition. Maize is also a good monocot model system for other possible bioenergy grass crops such as switchgrass and *Miscanthus* (de Leon *et al.*). Recently, hemp has been considered as a bioenergy source because studies have shown that it could be economical option and significantly mitigates GHG emissions especially when compared to *Miscanthus* and willow. It produces high yields of biomass with no agrochemical input and very little fertilizer use and it offers the potential of being an effective break crop as well as an energy crop (Finnan *et al.*).

The attitudes and positions held by the public and by key stakeholders are important considerations when discussing sustainable energy potential. Continuous dialogue is important to understand the views and opinions of people, groups, and other stakeholders that play important roles in further developing sustainable bioenergy contributions. The Roundtable on Sustainable Biomaterials (RSB), founded in 2007 and based in Geneva, Switzerland, is an international initiative involving various stakeholders that brings together farmers, companies, non-governmental organizations, experts, governments, and inter-governmental agencies that promote and advocate bioenergy sustainability production and processing. The RSB's mission is to "provide and promote the global standard for socially, environmentally, and economically sustainable production and conversion of biomass, to provide a global platform for multi-stakeholder dialogue and consensus building, to ensure that users and producers have access to credible, practical, and affordable certification, and to support continuous improvement through application of the standard" (RSB, 2013). In 2008, the RBS announced a dozen standards regarding sustainable biofuels, the most important being the following: biofuel production must adhere to international treaties and national laws, biofuel project planning will include relevant stakeholders, biofuels must mitigate GHG emissions, production cannot violate labor or human right laws, production cannot damage or threaten the food supply, biofuels must be produced in a cost-efficient manner, and shall avoid adverse environmental effects.

As with all innovative rising technologies, biofuel production remains controversial in some areas due to the adverse effects it may have on food sources and various involved stakeholders. This chapter begins by discussing the hindrances involved with using agricultural crops and natural ecosystems as sources of biofuel, then moves on to discuss the evolving stakeholders' interest in the matter, along with positions held by farmers, specifically from Kansas. The chapter then moves to discuss studies done using specific potential biofuel crops such as *Miscanthus*, maize, and hemp, analyzing how cultivation and production techniques affect surrounding areas and how surrounding areas affect the process. The last section examines energy storage by season from producing bioenergy from abandoned croplands.

Shortcomings of Using Agricultural Crops and Natural Ecosystems as Energy Sources

Foreign oil dependency and growing climate change are currently growing concerns around the world and have contributed to an increasing interest in using biofuels as an alternative to fossil fuels such as coal, gas, and oil. The study conducted by Graeme I. Pearman (2013) demonstrates, however, that bio-fuels and bio-sequestration can only make a minor contribution to lowering atmosphere carbon levels and minimizing net emissions of carbon into the atmosphere. This is done through examining available solar radiation and observing how efficient natural and agricultural ecosystems are in converting that energy to usable biomass. The 11 countries compared in the study are Australia, Brazil, China, Japan, Republic of Korea, New Zealand, Papua New Guinea, Singapore, Sweden, United Kingdom, and United States, with a main focus on the researcher's homeland, Australia. The objective of the study is to answer the following question: from a biophysical perspective, can using bio-fuels or bio-sequestration of carbon significantly contribute to the future of energy and the reduction of greenhouse-gas (GHG) emissions?

The first part of the study focuses on comparing annual rates of solar radiation and respective energy consumption for each country. The results group countries into 3 groups. Group 1, Japan, Korea, and Singapore had energy consumption around 1—10% of incident (surface) radiation. Group 2, China, UK, and US had energy consumption around 0.1% and Group 3, Australia, Brazil, New Zealand, Papua New Guinea, and Sweden had energy consumption around 0.1—0.001% of incident radiation. These comparisons demonstrate the limits that deriving energy from the sun has on meeting national expectations for energy consumption. We can consume much more energy than the sun could ever provide us.

Photosynthetic efficiency is another limit to the use of bio-fuels or bio-sequestration. The pigments in the chloroplast are only activated by certain parts of the solar spectrum, leaving much of the solar radiation unutilized. In addition, more than 50% of photosynthetic products (sugars) are lost through photorespiration. The whole process is only 3.3% efficient in C3 plants and 6.7% in C4 plants.

The study then continues to examine the limitations of bio-fuels regarding energy efficiency captured from natural vegetation and from global crops. Net primary production (NPP) is how much carbon (or energy in this case) remains after the photosynthetic organism has used it for growth and other metabolic functions. In natural environments, a large portion of captured solar energy is used within the community and is vital for a functioning and healthy ecosystem. Thus, human use of this energy will no doubt have negative impacts on preexisting ecosystems. Agricultural ecosystems are constructed for the purpose of providing biomass for human consumption. The main difference between the two types of ecosystems is that a cultivated system inputs fossil fuels, which needs to be considered when accounting for the net production of energy. Comparisons within each of the countries were then made between energy captured annually as net primary production and the national solar radiation and energy consumption rates. The comparison demonstrates the inefficiency of the biochemistry involved in photosynthesis and is also influenced by temperature and water availability. The comparisons also conclude that modifying the NPP of the biosphere could be possible when global scale changes occur to temperature, rainfall, and carbon dioxide concentrations.

Photosynthesis can be more efficient in agricultural crops when there is plenty of water and fertilizer and crop management is most favorable during the peak growth rates. In the study, multiple samples were taken from various countries and locations in order to accurately compare the relative efficiencies of different cropping systems. This is called "tradable production" because the net production is calculated after discarding the roots, leaves, and stems of plants. Sugar cane and wheat crops have the potential to contribute significantly to the nation's energy demand, but have some economic and political shortcomings that are not discussed in detail in the paper.

Though natural and agricultural biomass have the potential to provide energy for human use and to offset carbon emissions from fossil fuels, this study demonstrates that there are major limiting factors to this solution including the availability of solar radiation and the efficiency of photosynthesis needed to convert the energy into feedstock. Another limitation is how efficiently biomass can be converted into fuels that are appropriate for existing feedstocks, conversion systems, and applications. Solar radiation on land accounts for 1700 times the amount of energy consumed by humans, but the radiation and the energy demands are not evenly distributed geographically, so this process depends on the redistribution of energy. It also depends on how efficiently solar energy can be converted to meet the demands of humans, which is where photosynthesis becomes a limiting factor. In comparison, agricultural crops may be more efficient at converting solar radiation to a more usable form of energy, but the study demonstrates that wheat, rice, and corn crops have low efficiency rates that are similar to those of natural ecosystems. The only crop that shows better efficiency is sugar cane.

The analysis conducted in this study is not meant to completely reject the idea of using crops and natural ecosystems for bio-fuel and bio-sequestration of carbon, but is meant to illustrate that this would require a huge amount of increase in land utilization and/or altering existing crops. Investors in these types of activities and governments seeking policy implementation need to be skeptical of these so-called "attractive" energy efficiency solutions.

The paper summarizes 12 issues raised by the possibility of using bio-fuels as a future energy source and for the bio-sequestration of carbon. The first issue is the potential for agricultural and forestry capacity to deliver to energy demands and emissions reduction. Another one is evaluating the co-benefits or dis-benefits of developing policies about bio-fuels such as soil productivity, job creation, economic opportunities, international balance of trade, security of energy supply and so on. There also has to be enough net energy to cultivate crops for fuels, to produce fertilizer, transform the energy into chemical energy, and for transporting the subsequent fuel Another issue to keep in mind is the continuously changing climate and its effect on which bio-fuels are appropriate. Other issues include timing, production location, strategic carbon and nitrogen budgeting, human capacity to convert the energy, competing use of land, costs of production, and social and political realities.

The conclusion of the paper does little to provide the answers to the various questions raised throughout the study, but rather implies that "we" have the knowledge to develop a system to produce bio-fuels and bio-sequestration of carbon from agricultural crops and natural ecosystems, but now we need more efficient biomass that will provide us with the tools we need to power that process.

Advancing Sustainable Bioenergy: Evolving Stakeholder Interests and the Relevance of Research

The diversity of stakeholders' interest and values complicates the decision-making process involved in the future of sustainable bioenergy production. Johnson *et al.* explores the different stakeholder perspectives and then examines how this diversity affects research on the subject. Biofuel production has been brought to the public's attention because of the need to mitigate greenhouse gas (GHG) emissions, increase energy security, support farm production, and improve economic growth in rural areas. The recent increase in biofuel consumption has resulted in stakeholders questioning environmental, economic, and social benefits of using agriculture to produce ethanol and biodiesel. As a result, policy makers have passed legislation and modified regulations about renewable fuel production in order to promote the use of alternative biomass feedstocks. The general research community is looking for ways to convert this feedstock to a usable fuel source in vehicles. The expansion of biofuel production coincides with the addition of more and diverse stakeholder groups that will be involved or affected by various parts of the biofuel supply chain. These groups require different information to make decisions and have conflicting perspectives about sustainable bioenergy. Johnson *et al.* investigate the implications of stakeholder diversity for bioenergy research and inform research communities the importance of social science issues in relation to their work.

Sustainable bioenergy production is defined as the use of biomass as an energy resource that contributes to climate change mitigation, energy security, and economic development goals, results in minimal environmental and social impacts, and attains economic self-sufficiency. Differentiating stakeholder interests results from varying values and objectives and leads to fragmented decision-making. Although the United States Renewable Fuel Standard revisions are making progress toward mitigating ineffective decision-making, research alone will not resolve existing issues. Increasing use of biomass as energy feedstock has led researchers and environmental community members to investigate the environmental, social, and economic impacts of using bioenergy in a way that would mitigate our petroleum consumption. The concerns of these groups include the net energy balance of biofuel production, climate benefits, ecosystem service impacts, global impacts on international land use, agricultural markets, and food security. These concerns relate to the current limited range of biomass feedstock and the hope of developing "second generation" cellulosic feedstock that includes more than corn grain and soy. However, the technology developments for this endeavor are uncertain.

Stakeholder diversity results from the complexity of bioenergy production: it's an industry that is not vertically integrated. The supply chain intersects with economic, technical, and regulatory systems. Stakeholder diversity also stems from the uncertainty of bioenergy production because many feedstocks lack successful markets, conversion technology is not perfected or widely available, and funding is limited. Stakeholders are defined as groups that participate in or are affected by public and private decision-making with regard to bioenergy production. Stakeholders are either directly involved with the bioenergy supply chain, affected by bioenergy production indirectly, responsible for governance of the supply chain through developing policies and regulations, or are interested in advancing the development of the bioenergy industry.

Researchers should be aware of how differing stakeholder groups will affect the questions they ask and the information they provide. Being aware of the differences will lead to sustainable research outcomes. A prominent issue is that as new stakeholders arise, so do new issues and these new issues attract new stakeholders and continue the complex cycle resulting in intricate research dilemmas. Researchers can affect this issue domain by trying to reconcile doubts, adding a level of understanding to complex issues, and increasing or decreasing the relevance of specific subjects. For example, there is no simple answer to whether biomass as an energy resource will be beneficial because it relies on stakeholder perspective and values. Thus, researchers need to begin their work by studying and understanding the full range of stakeholder values.

The success of sustainable bioenergy production depends on scientific and technical information that align with stakeholder values and interests. One of the barriers to this production is the lack of processes to negotiate effectively among competing stakeholder groups. In order to make the bioenergy successful, stakeholders need to engage with one another or else they risk myopic decision-making. For example, conflicting values of international land use impede the development of the bioenergy industry due to the uncertainty about which resources should be used as feedstocks for renewable fuels.

Some argue that policy and regulation could alleviate these types of conflicts, but the continuously changing circumstances prevent these regulations from being flexible and effective. Stakeholder diversity also affects the type of information that researchers provide. Three characteristics that increase research influence are saliency, credibility, and legitimacy. Stakeholders evaluate information against these characteristics, but will weigh each characteristic differently. Thus researchers need to frame their work in order to make it appealing to the particular groups they are targeting. Because one information source cannot maximize all three characteristics simultaneously, a large range of information resources is needed.

Johnson *et al.* discuss various recommendations for the research community. The first recommendation is concerned with the value of participatory research—stakeholders need to be active participants in research activities because they can improve research design, enhance decision support, and increase vital attributes of research output. However, participatory research does not imply favoritism of a specific stakeholder group. The second recommendation suggests that the bioenergy research community should focus on improving the interactions among stakeholders. Lastly, the authors recommend rigorous and continuous bioenergy stakeholder analysis in order to understand changing values, interests, and decision-making processes.

Shifting Lands: Exploring Kansas Farmer Decision-Making in an Era of Climate Change and Biofuels Production

Using Kansas as an example, White *et al.* examine the various factors that influence farmer decision-making during this controversial era of climate change and energy conservation. A conceptual model for understanding farmers' decisions was developed from interviews conducted with a diversity of farmers and key informants. Interestingly enough, the results demonstrate that most farmers hold a positive perception of the natural environment and don't have a strong concern about climate change issues. The guiding factors of farmer's decisions about whether or not to cultivate biofuel crops are the relative advantages of the practice and the ability to discuss the practice with a social network. There is a strong need to create a renewable energy market in the US because of its potential to reduce greenhouse gases and increase production benefits; biofuel crops pose one plausible solution. The paper addresses the following question: considering global climate and energy concerns, what are the main influences on farmers' decisions regarding land use, specifically the decision to cultivate biofuel crops?

Kansas is one of the least populated states in the US due to its rural, agricultural character. In recent years it has experienced a decline in the number of farms, but an increase in farm size. Farmers are less concerned with planning for future climate change, than with more specific and prominent issues such as water supply. However, the state has taken an active role in discussing the potential benefits and use of biofuels, especially ethanol. Kansas currently has 11 ethanol plants and 3 biodiesel plants. But the state's uncertain future about land use and agriculture due to dramatic climate changes has led scientists to predict that Kansas farming is likely to experience significant disruption and adjustment.

As climate change begins to threaten the future of agriculture, it's important to understand what drives farmer behavior and decision-making. Recent literature examines factors such as motives, values, and attitudes that construct farmer management and decisions regarding conservation. The studies conclude that farmers prefer practices that don't hinder farm productivity and that relate to the perception of a "good farm" landscape. Furthermore, studies show that farmers usually make production and management decisions using their own local knowledge. When making decisions about growing biofuel crops, farmers' biggest concerns were visual impact of the land and the risks associated with growing such crops. A study conducted a few years back reveals that younger farmers with higher income and higher education levels were more willing to convert part of their land to bioenergy crops. Another interview-related experiment demonstrates that farmers are skeptical about the economic benefits of biofuels and unsure about the success of an agricultural bio-economy. Farmers are most concerned about crop profitability, the cost of switching to new crops, and the incompatibility of growing biofuel crops with traditional ones.

The connection between farmers and their communities is also another important factor that affects decision-making about biofuel crops. Research shows that corporate-style farms had a negative impact on the unity of the community, leading to less civic engagement. The community also has a strong influence on farmer decision-making through face-to-face communication, local social networks, cultural and social norms, and local support structures.

The study conducted two phases of semi-structured interviews. Sixteen interviews were conducted with key informants with expertise in environmental issues and the other 17 interviews were conducted on a diverse group of Kansas farmers. The interview questions were tailored according to the expertise of the interviewees. Farmers were interviewed in the following topics: farm decision and plans, new practices, programs and policies, weather/climate influence, and attitudes regarding biofuel crops.

The interviews with the key informants demonstrated that organizations and agencies that work closely with farmers view farmers' decisions as multifaceted and difficult to understand. However, several common themes did emerge from the interviews. One key theme informants found was that farmers make agricultural decisions with the most regard toward potential economic benefit. Farmers investigate advantages that new practices will bring and observe effects of these new practices in other places before implementing them in their own farms. Another factor that informants saw as influential in farmers' decisions was related to the characteristics of the decision setting such as the presence of biofuels processing facilities and farm policies. Informants also observed good stewardship of the farmers as another factor influencing decision-making.

When interviewing the farmers, it became clear that the farmers had a skewed perception about the changing environment and climate. They considered short-run weather changes such as extreme heat or cold, but did not consider long-run impacts of climate change, indicating that a farmer would most likely not consider climate change when making innovative agricultural decisions. The farmers' interviews also revealed that there had to be clear benefits and no risks when changing agricultural practices. The most important factors to them were profitability, lack of risk, efficiency, and environmental benefits. Thus, any new practice such as the cultivation of biocrops on existing land, needs to provide a benefit over the existing cultivation or farmers will not even consider making the switch. Important characteristics of the interviewed farmers include openness to new ideas, curiosity about existing practices, an emphasis on productivity and stewardshipness, a desire to learn about new practices, and experience or knowledge with biofuel crops.

Another aspect of the study revealed the various characteristics of the decision setting: local social networks and broad cultural attitudes, agricultural policies and programs, and social institution. Social networks and broad cultural attitudes are an important part of farmers' relationships with communities. According to farmers, other local farmers are their most important source of information about new farming techniques and practices. It's a resource for continual support. Another important part of social networks is the basic quality of life in rural areas. The interviews revealed two different perceptions regarding agricultural policies amongst the key informants and the farmers. When asked what policies would most likely influence farmer decision-making, key informants mentioned a wide variety of federal bills whereas farmers were not well informed at all about these policies and programs. Many farmers expressed conflicted views of government incentives and subsidies for farming activities.

The amount of data collected from the interviews reveals the complexity of farmers' decision-making. It can be concluded that farmers currently have a positive perception of the natural environment and do not appear to have legitimate concerns about climate change. However, decisions are greatly influenced by farm practices—that is a new practice must have clear and tangible advantages over an already existing practice. The interviewed farmers viewed themselves as seeking information, but not innovation, meaning that farmers are open to new ideas, but don't want to be the first ones to try something new. The study does little to draw any conclusions about the future of agricultural production in Kansas due to the limited farmer sample and collected data. The main conclusion resulting from this study is that agricultural practices that are most likely to resonate with Kansas farmers are those that align with local environmental conditions, provide clear advantages, can easily be learned about through existing social networks, and provide farmers with a sense of independence and contribution to the good of the community. Factors that may have a negative impact of farmer decision-making include incentives and subsidies of new activities through government policies or appeals to mitigate climate change. More research on a larger scale is needed to draw any further conclusions about the future of biofuel cultivation in Kansas.

Saving Land and Water by Cultivating *Miscanthus*

Due to government mandates in response to climate change, ethanol production has steeply increased since 2009, and there are now for 79 billion liters of cellulosic biofuels yearly by 2022. Cellulosic crops such as maize, switch grass, and *Miscanthus* have been determined to be viable biofuel sources. In order to meet the biofuel target in 2022, cellulosic crop cultivation needs to be expanded and intensified. The impact on land and water use needs to be considered as well. Zhuang *et al.* present a data-model assimilation analysis assuming that maize, switchgrass, and *Miscanthus* can be grown on available U.S. croplands.

The current production levels of maize are not enough to be simultaneously used as biofuels and as a food source. The cellulosic crops switchgrass and *Miscanthus* have been identified as viable alternatives to maize in producing second-generation biofuel. This is staged to work especially well in temperate regions because of their higher biomass productivity and available crop-producing land. Other studies have shown that bioenergy crops have higher land and water efficiencies than food crops do, but the increasing demand of land and water to cultivate these crops hasn't been researched using ecosystem models. The study uses the terrestrial ecosystem model (TEM) to predict the demand of land and water for growing various biofuel crops so that enough ethanol can be produced to hit the 2022 target. The goal of the study is to analyze the demand for resources rather than to analyze the environmental impact of growing biofuel crops.

The TEM ecosystem model uses gross primary production (GPP) as the core algorithm, which describes the rate at which a plant produces usable chemical energy. The primary production (NPP) is the difference between

GPP and plant respiration. In order to analyze the productivity of feedstocks and biofuels, the researchers estimated the biomass and biofuel production in terms of harvestable biomass (HBIO) and bioethanol yield. Current and future biofuel production was estimated using conversion efficiencies and currently available and potentially advanced technologies. TEM was run several times at each site in order to achieve model equilibrium. Analyses were conducted on biomass and biofuel yield, water balance, and water use efficiency and were estimated based on simulations.

The results of the model demonstrate that in order to produce 79 billion liters of ethanol from maize grain, there would be a need for 190 million tons of conventional grain and 26.5 million hectares of land, which is equivalent to 20% of total US cropland. The water loss of this production would be 92 km^3, but if the maize stover were also used, water would be saved. Because switchgrass has lower conversion efficiency, using this crop would result in a higher demand of biomass. More land and water would need to be used to produce the same amount of ethanol than using maize. Alternatively, *Miscanthus* would only require half the amount of land and two-thirds the amount of water used for maize grain in order to produce the same amount of ethanol. Furthermore, with the advanced technologies predicted for future years, even less water and land will need to be used in converting biomass to biofuel. The model experiments demonstrate that switching from maize to *Miscanthus* will save land and water, but that switching from maize to switchgrass will require more land and water.

This study only predicts ethanol production using available croplands, but recent studies have illustrated that marginal lands could also be a source for cultivating cellulosic crops. Experiments have also shown that switchgrass may be more productive on marginal lands than on traditional croplands. The model may produce some bias because it does not consider the effects of fertilization, irrigation, rotation, and tillage. To strengthen the study, analysis of economic viability, food security, nutritional and ethical concerns, and other environmental consequences and benefits need to be conducted.

The Critical Role of Extreme Heat for Maize Production in the United States

Maize production continues to be a very important source of food, feed, and fuel all around the world, but climate change has raised the concern about being able to maintain the yield rates. A negative relationship between extremely high temperatures (above 30°C) and yield has already been observed in various regions. Previous studies have not been able to demonstrate which mechanism causes the correlation between extreme temperatures and yield, thus it is possible that the relationship reflects the influence of another variable, such as precipitation rates. There are other possible explanations for the observed relationships. This study explores the mechanisms used in other studies that document the importance of extreme heat on rainfed maize using the process-based Agricultural Production Systems Simulator (APSIM). The study asks three main questions: can APSIM reproduce the empirical relationships?; if so, what does APSIM imply are the key processes that give rise to these relationships?; how much are these relationships affected by changes in atmospheric CO_2?

Simulations by APSIM in Iowa demonstrated a strong and significant relationship between yield and extreme degree days (EDD). This result was observed in other simulations as well, implying the importance of EDD in previous studies. The agreement between APSIM and empirical relationships also implies that it is unlikely that the effects of heat on flowering or respiration rates are related to the relationships because APSIM does not model these effects. Another implication is that the lower importance of total rainfall observed in previous studies may merely be a feature of maize productivity in the evaluated region. The simulations also demonstrated a weak relationship between crop biomass accumulation at the daily timescale and daytime maximum high temperature (T). This result indicates that the direct effects of T on net photosynthetic rates in APSIM cannot explain the importance of extreme heat; rather daily growth is more associated with the water stress index (soil water supply/soil water demand). July had the most biomass growth, indicating that water stress during this month is particularly important for yield rates.

High temperatures influence water stress through their effects on water demand operating through vapor pressure deficit (VPD). The demand for water doubles as the temperature increases from 27°C to 35°C. On the other hand, by lowering transpiration efficiency (TE), high T causes more soil water loss, reducing the future values of the soil water content and the water supply. Thus, the relationship between T and water supply differs on daily and monthly timescales. Another simulation was run to compare the influence of T on precipitation (P). The results show that raising T increased water demand in every month and reduced water supply. Large P changes are necessary to offset the effect of T on water stress because high T affects water demand and supply.

The EDD-yield relationship was further tested by calibrating a model to simulations with historical temperatures and then predicting scenarios. Agreement between yield changes in APSIM and this statistical model indicated that EDD is directly associated with low yields through the link of EDD to increased VPD. An important factor that influences water stress in APSIM is the TE coefficient (TE_c) because of how it affects water demand. By improving this coefficient in C3 and C4 crops, increases in CO_2 concentrations will reduce sensitivity of yields to EDD. Although increased CO_2 levels may reduce EDD sensitivity, there are other factors that may be increasing it. The reduced sensitivity could potentially be counteracted by trends in crop improvements favored in cooler temperatures. More research is needed to understand the interactions between genetic improvements, drought, and CO_2.

Correlations between EDD and maize yields have been successfully reproduced by APSIM simulations. The overall results of this simulation suggest a minor role for direct heat stress on reproductive organs at current temperatures in the simulated region and that the effects of increasing CO_2 concentrations on transpiration efficiency should reduce yield sensitivity to EDD in upcoming decades.

Hemp: A More Sustainable Annual Energy Crop for Climate and Energy Policy

Growing concern about greenhouse gas (GHG) emissions and climate change due to fossil fuel dependency has led to the consideration of more attractive energy sources, especially bioenergy sources. In Northern Europe, the two crops that have worked the most effectively are *Miscanthus* and willow, two perennial energy grasses that have proven to be sustainable energy crops due to high yields of biomass from low inputs. Farmers are fairly attracted to cultivating these crops because of the declining farming market and the future promise of a biomass energy market. Farmers use break crops to control for disease and weeds, a technique that also increases wheat production. Currently sugar beet and oilseed rape are used in Northern Europe, but because of reduction in the sugar beet industry, hemp has been predicted to be an effective break crop because its root system aids soil structure. Various studies have demonstrated that it produces high yields of biomass with no agrochemical input and very little fertilizer use. It offers the potential of being an effective break crop as well as an energy crop. Finnan *et al.* compare hemp with other annual and perennial energy crops, economically and as a way to mitigate GHG emissions.

Hemp was compared with sugar beet and oilseed rape using Life Cycle Assessment (LCA) and Net Present Value (NPV) economic assessment on collected Irish data. The environmental and economic comparison requires accurate yield estimates, which vary as a result of meteorological conditions, agronomic practices, and soil type. Because of this the crops were considered across four different levels: a low yield, two mid-level yields, and a high yield. Hemp, sugar beet, and oilseed rape were all grown on tillage farms as break crops. The ground was prepared for *Miscanthus* cultivation by applying herbicide and then subsoiling and ploughing. Average farm models were built using previous models as examples. Inputs and sinks of GHG's were considered. Energy use was divided into two categories: those that used diesel and those which used electricity.

Carbon is stored under roots and stays in the soil for long periods of time, thus increasing carbon levels in soils correlates to long-term removal from the atmosphere. Other studies have shown that crop rotations increase soil carbon. In this study two scenarios were considered for below ground carbon storage from perennial crops, both grassland and arable. The assumption that there would be no soil carbon increase when grasslands were converted to perennial energy crops was made, but that soil carbon would increase if perennial energy crops were sown on arable land. Net cultivation emissions were calculated by subtracting carbon sequestration from cultivation emissions. There are large land areas in the European Union (EU) used for liquid biofuel production from oilseed rape and sugar beet. Hemp could be grown on parts of this land to produce feedstock for heat and electricity. An economic analysis was then performed on the yield levels for each crop, assessing the life cycle of each crop.

The results of the study demonstrate that hemp's GHG mitigation potential is comparable to that of perennial energy crops such as *Miscanthus* and SRC grown on grassland. However, the results depend on the various assumptions, particularly the comparative yields and end use of biomass. Biomass production depends on local conditions such as climate and soil type. Productive lifetimes of perennial energy crops are unknown because it is hard to predict stresses that these crops will face over 15-20 year plantation lifetimes. Perennial energy crops are characterized by low cultivation inputs and by their potential for soil carbon sequestration, which is an effect that is most significant in tillage soils because of the low carbon content. Converting grassland to perennial crop energy has an initial expected loss of stored carbon following ploughing and soil preparation, but after the initial loss, soil carbon reserves return to greater levels than those of grassland soils. Therefore, low cultivation emissions and carbon sequestration ability are the advantages that perennial energy crops have over annual energy crops. Cultivation emissions are small compared to GHG mitigation through fuel substitution, thus hemp is favorable compared to perennial crops with regard to GHG reduction, assuming that all the crops studied have similar yield potential.

Compared to perennial energy crops, hemp has higher annual costs because of soil preparation, seed purchasing, and higher fertilizer requirements. However, hemp is more appealing to risk-averse farmers. Farmers would receive full returns within the year of planting and can continue or discontinue using hemp cultivation the next year based on the experience. In comparison, growing perennial energy crops requires a high initial investment, a long waiting period before cash flows are positive, and a commitment of land for over 20 years.

This study assumed that all of the compared crops would substitute for oil, but *Miscanthus* and hemp biomass would more likely be used for electricity generation through co-firing in coal and peat power stations in Ireland, which would require less processing and lead to more GHG abatement. Hemp could also substitute for a greater amount of oil than biodiesel and bioethanol from sugar beet. One major criticism of using annual energy crops for biofuel production is the detrimental impact on food supply. Thus, hemp is particularly attractive because it is not a food crop and acts as a low input break crop that improves soil quality and yields of subsequent crops. Hemp production can complement food production rather than compete with it. Another advantage of hemp over perennial energy crops is that it can be supplied immediately because it produces high biomass without having to wait until the end of the yield-building phase. Hemp may be an important crop to farmers because of the diminishing of the EU sugar sector in the last few years meaning that much of the tillage land in Europe does not have efficient break crops. Hemp is a better alternative to sugar beet as a break crop that can be used for bioenergy production and GHG mitigation.

Topographic and Soil Influences on Root Productivity of Three Bioenergy Cropping Systems

Root production in plants plays a vital role in ecosystem carbon, nutrient, and water cycling, but researchers have not made much progress in further understanding this issue. It's important to understand the impacts of environmental conditions on root production because it aids in the development of a sustainable bioeconomy. However, scaling root productivity estimates for cropping systems beyond plot scales poses a great challenge to researchers. Whether the bioenergy plants are annual or perennial influences the biogeochemical cycling and the ecological benefit of the systems. The foundation of the study is based on previous research of the response of root growth to variations in soil properties at multiple spatial scales. Roots of plants generally respond to different soil types by growing into nutrient patches, but this depends on the species and nutrient demands or limitations. Ontl *et al.* measured the response of root productivity of three different bioenergy cropping systems across a topographic gradient with variation in typical agroecosystem soil conditions. The hypothesis is that root dynamics would vary by cropping system and position of the landscape across a hillslope. If landscape alone was not a good enough indicator, they predicted that root productivity would be related to differences in soil.

Three cropping systems were observed: switchgrass, continuous corn, and triticale/sorghum double crop. The annual root production was measured by evaluating various soil properties as indicators of soil functions. The results demonstrate that the annual root production of each cropping system was significantly different, implying that bioenergy cropping systems are efficient indicators for scaling estimates of root production to landscapes. Switchgrass had produced twice as much root biomass as continuous corn and triticale/sorghum double crop, demonstrating a significant difference in the cropping systems. Furthermore triticale/sorghum produced a significantly greater amount of root biomass than continuous corn. The results also demonstrate that landscape position was a poor predictor of root production because the switchgrass' influence. There was no significant effect of landscape position on root production. However, an analysis on the 11 soil parameters demonstrated a significant difference in landscape positions, but did not strongly support the hypothesis that landscape position is a useful indicator of root productivity, nor are the differences in soil properties. The results do support the alternative hypothesis that root production can be estimated based on differences in edaphic characteristics. The results also suggest that the percent of sand in the soil can serve as a substitute for other important soil variables.

The study provides evidence that annual and perennial bioenergy cropping systems differ in root productivity, and also that heterogeneous edaphic conditions differently impact root production of cropping systems. The differences in root production of the three cropping systems demonstrate the strength of bioenergy crops as predictors of root productivity. The root productivity of the annual bioenergy cropping systems wasn't affected by landscape position; however, the root production in switchgrass was lower in the floodplains suggesting that landscape position could potentially have some predictive value. Further studies and empirical evidence is needed to understand and test the study's findings on a broader level.

Seasonal Energy Storage using Bioenergy Production from Abandoned Croplands

Producing electricity from biomass could potentially provide a back-up storage source for the intermittency that accompanies wind and solar energy production. Biomass electricity also provides a carbon-negative and efficient method for bioenergy production, which is important because of mandated restrictions on carbon emissions. Furthermore, biomass electricity also provides an efficient method for providing renewable transportation energy that could replace current liquid fuel approaches. Although bioenergy may be important in producing electricity and developing energy storage mechanisms, the economic and environmental effects are unclear. Studies have been conducted on abandoned agricultural lands to try to find a path of producing bioenergy that has reduced land impact. Campbell *et al.* estimate at county level, the magnitude and distribution of abandoned agricultural lands in the United States and attempt to quantify how much potential energy storage could be produced by the provided bioenergy.

GIS-modeling was used to develop maps of availability, biomass yields, and bioenergy production of the US counties studied. Abandoned agricultural lands were further divided into abandoned croplands and abandoned pastures. The new cropland estimates were compared to previous estimates based on data from other global gridded databases. Seasonal energy storage requirements were estimated using data from current electricity demand, wind production, and solar production from the Department of Energy's (DOE) Energy Information Administration (EIA). Annual storage requirement is the storage capacity that is required to counterbalance energy deficits that result from seasonal intermittency. The current abandoned cropland estimate is 71 million hectares (Mha), which is about 41% of current cropland area.

The high-resolution land-use databases provided a 61% larger area estimate of abandoned croplands than previous studies reported. This large estimate resulted from using more spatially resolved data. The results also suggest that there is a larger uncertainty in abandoned agriculture than previous studies concluded. Future analysis and studies of bioenergy potential should consider availability of land resources, yield estimates, and energy conversion efficiencies. The collected data were used to consider the role of bioenergy in seasonal energy storage, which is necessary because of the intermittency of future energy production based on wind and solar energy. The results demonstrate that bioenergy could provide most seasonal storage requirements, but that a system dominated by solar energy requires positive assumptions about biomass yields to completely satisfy the energy storage requirement.

Conclusions

The implications that sustainable bioenergy could have in transforming and revolutionizing the fuel industry are important when discussing the future of the world's energy sources. As we deplete natural resources, we will inevitably need to convert to innovative new ways of producing energy sources, and bioenergy appears to be the most viable and obtainable option. Bioenergy production will mitigate GHG emissions and provide economic benefits, however there are also potential limitations that have been studied. Some of these limitations include the availability of solar radiation and the efficiency of photosynthesis needed to convert the energy into feedstock, along with how efficiently biomass can be converted into fuels that are appropriate for existing feedstocks, conversion systems, and applications. Furthermore, the various interests held by stakeholders affect the decision-making process regarding biofuels production. It is a complicated process because the bioenergy industry is not vertically integrated—it is a wide spectrum. Farmers play a prominent and important role in biofuel production and it is essential to understand the decision-making process from their point of view. Studies have shown potential benefits of using biofuel crops, such as saving land and water by cultivating *Miscanthus* and enhancing root productivity by using different cropping systems. Another study demonstrated how bioenergy production from abandoned croplands could be used in energy storage.

References Cited

Brosse, N., Dufour, A., Meng, X., Sun, Q., Ragauskas, A. 2012. Miscanthus: a fast-growing crop for biofuels and chemical production. Bioreference 6, 580—598.

Campbell, J., Lobell, D., Genova, R., Zumkehr, A., Field, C. 2013. Seasonal energy storage using bioenergy production from abandoned croplands. Environmental Research Letters 8, 1—7.

De Leon, N., Kaeppler, S., Lauer, J. 2013. Breeding maize for lignocellulosic biofuel production. Bioenergy Feedstocks: Breeding and Genetics, 8 151—167.

Finnan, J., Styles, D. 2013. Hemp: a more sustainable annual energy crop for climate and energy policy. Energy policy 58, 152—162.

Johnson, T., Bielicki, J., Dodder, R., Hilliard, M., Kaplan, O., Miller, A. 2013. Advancing sustainable bioenergy: evolving stakeholder interests and the relevance of research. Environmental Management, 51: 339-353.

Lobell, D., Hammer, G., McLean, G., Messina, C., Roberts, M., Schlenker, W. 2013. The critical role of extreme heat for maize production in the United States. Nature Climate Change 3, 497-501.

Ontl, T., Hofmockel, K., Cambardella, C., Schlute, L., Kolka, R. 2013. Topographic and soil influences on root productivity of three bioenergy cropping systems. New Phytologist 199, 227-737.

Pearman, G. 2013. Limits to the potential of bio-fuels and bio-sequestration of carbon. Energy Policy 59, 523-535.

White, S., Selfa, T., 2012. Shifting lands: exploring Kansas farmer decision-making in an era of climate change and biofuels production. Environmental Management 51, 379—391.

Zhuang, Q. Qin, Z. Chen, M. 2013. Biofuel, land, and water: maize, switchgrass, or *Miscanthus*. Environ. Res. Lett. 8, 015020.

Roundtable for Sustainable Bioenergy. 2013. Vision and mission statement for the RSB. Web. 6 . Nov. 2013. http://rsb.org

Sustainable Biofuels. Wikipedia: the Free Encyclopedia. Wikimedia Foundation, Inc. 7 Oct. 2013. Web. 5 Nov. 2013. http://en.wikipedia.org/wiki/Sustainable_biofuel

.

3. Smart Grids to Reduce Carbon Emissions from the Energy Sector

Stephanie Oehler

The electric grid dates back to the 20th century, when its primary purpose was simply to deliver power to customers. Over time, however, inefficiencies and areas for improvement in the generation and distribution of energy have become increasingly evident. Arnold (2011) identified the primary incentives for updating the grid in the modern era, which include: improving cost-effectiveness in production and transmission; allowing customers to play more of an active role in power purchases through enhanced information services; incorporating renewable energy sources to reduce impacts to global climate change; increasing reliability; and growing the capacity of the grid to incorporate devices that will be progressively more dependent on electricity as fossil fuels are phased out of use. The author believed that these goals should be accomplished through improved technological features that monitor various aspects of the grid, reduction of peak demand through demand-side management measures, and the incorporation of renewable energy sources (Arnold, 2011). These measures have the ability to transform the rigid and outdated electricity system into the modern-day smart grid.

In 2013, the Intergovernmental Panel on Climate Change (IPCC) constructed climate models and found that observed variations in climate resulted from a combination of natural and human-based factors. Climate models were used to simulate future conditions based upon adjusted variables in order to determine which modifications caused the changes the world has already experienced and to predict the long-term impacts they will have on climate. The IPCC models supported the theory that climate changes observed today are predominantly the result of human behavior, particularly since human-driven, industrial greenhouse gas emissions account for more than 50% of the total global concentration. Atmospheric carbon dioxide concentrations, which are most directly influenced by the combustion of fossil fuels, have increased 40% since 1750. While continued progression under these conditions will have varying impacts depending on location, extreme surface temperatures, ocean temperature warming, unpredictable rain events, sea level rises, and sea ice melting are among the predicted outcomes (IPCC, 2013).

A considerable amount of carbon dioxide emissions are attributed to fossil fuel combustion and can be traced, primarily, back to energy production and transmission systems. While the energy sector is responsible for a large share of the atmospheric greenhouse gas concentrations that are driving climate change, studies have shown that the impacts of increased carbon in the atmosphere pose a tremendous threat to the stability of the electricity grid of the future (Lyster, 2013). Efforts to improve the efficiency of the grid, thus, are two-fold in purpose. In bolstering systems to increase their resistance against extreme and unpredictable weather events, developers also seek to reduce the impacts of the grid on climate change in the future(Guccione, 2013). These issues need to be addressed through multiple efforts, including technological, policy, and market improvements, in order to bring about the significant changes that are required for the security and durability of energy supplies and climate stability (US Department of Energy, 2013).

There are many aspects of the electricity grid that require improvements. Guccione (2013) discussed the ability of smart grid solutions to both mitigate and adapt to climate change through gains in efficiency. Reducing demand, implementing more efficient technologies, and incorporating renewable sources allows for increased interaction between energy producers and consumers, cost savings, emissions reductions, and improved sustainability of the grid. When complimented by policy changes and market-incentives, technologies and management changes are realistic options that the demand and supply-sides of the energy sector can implement in order to combat the increasing role of electricity generation in perpetuating climate change through carbon emissions.

Concisely defining the smart grid is a challenging endeavor. At its most basic level, a smart grid incorporates technologies that reduce inefficiencies in the production and transmission of electricity, improves monitoring and communicating capacities between consumers and producers, allows for the integration of cleaner, often renewable energy sources, and supports conservation measures amongst both users and suppliers of energy. Characteristics and capacities of smart grids vary by location depending on the existing energy profile, grid capacity, types of consumers, access to resources, and market and policy conditions. As a result, a variety of approaches have been taken to transform outdated, inefficient, and unreliable electricity grids into smarter systems, each with varying degrees of success. Technological developments continue to improve the ability of the grid to be self-sustaining, interactive, and inclusive of cleaner types of energy. Existing grids face challenges in integrating renewable energy sources, largely due to their unpredictable supplies which are dependent on irregular conditions such as sun and wind. In addition to reducing costs for consumers and utilities, smart grids can be used to reduce greenhouse gas emissions by improving efficiency and incorporating energy sources that have significantly fewer or no emissions, as compared to fossil fuels. By using technology to manage multiple energy sources and provide vessels for storage, such as electric vehicles, smart grids utilize creative methods to continue to meet the ever-increasing demand for energy.

Scholars have also examined the potential of technologies to collect data that can influence consumer demand, alter the dynamics between consumers and producers, and increase predictability and ease of managing demand levels to minimize reliance on back-up generation methods. A sizeable percentage of greenhouse gas emissions resulting from energy production stem from the generators that are relied upon during peak demand, when other lower emitting sources are not providing an adequate supply. While renewable energy sources offer one solution to minimize this unsavory production, certain technologies, partnered with new energy policies and market incentives, have the potential to alter consumer demand flexibility which would reduce such peaks in the overall demand curve. Upgrades to grid technology combined with increased interactions between consumers and the grid have the potential to improve energy efficiency and reduce reliance on fossil fuels by increasing the predictability of demand.

A number of these approaches have been implemented throughout the world, particularly in developed nations. The United States passed the Energy Independence and Security Act (EISA) in 2007, offering support to the existing smart grid efforts and requesting an allocation of funds to establish and support a Smart Grid Advisory Committee, the Federal Smart Grid Task Force, the "Smart Grid Regional Demonstration Initiative," a Smart Grid Interoperability Framework, and a "Federal Matching Fund for Smart Grid Investment Costs." According to the US Department of Energy's Office of Electricity Delivery and Energy Reliability, with more than 12 additional federal departments and agencies involved with smart grid implementation and regulation in the United States, the Task Force seeks to integrate its efforts where practical and useful (US Department of Energy). One significant way in which the federal government supports efforts to develop smart grids is through the public-private partnerships that are formed through the Smart Grid Investment Grand Program, which have resulted in a $7.8 billion investment in 99 projects selected from across the country (US Department of Energy, 2012). The United States is not alone in its ambitious investment patterns, with countries across Europe, China, Japan, Korea, and Australia also offering financial support for grid modernization (Arnold, 2011).

In 2013, scholars explored the status of smart grid implementation, the successes of demand-side and supply-side management efforts, and impacts of specific technological advancements. Additionally, they sought to determine which methods were most effective in reducing carbon emissions and explored the impacts of policy and market-based solutions. This chapter provides a detailed overview of a variety of recent and relevant research in smart grid development and explores the potential of greenhouse gas emission reductions from a variety of grid modernization efforts.

Implementation Status of Electric Distribution Systems in US Smart Grid Projects Funded Under the 2009 American Recovery and Reinvestment Act

As climate change continues to have a greater impact on individuals and habitats around the world, many nations are taking actions to reduce their greenhouse gas emissions. Modernization of electrical grids in order to increase efficiency, accommodate new sources of energy and technology, minimize losses, and ultimately reduce harmful emissions from high-polluting forms of energy production has become a priority for many countries. For example, the American and Chinese governments each contributed over seven billion US dollars to national smart grid deployment in 2010, with numerous other developed countries investing similarly large amounts in their own electricity infrastructures. Ghosh et al. (2013) explored the current status of projects partially funded through the Smart Grid Investment Grant (SGIG) and the Smart Grid Demonstration (SGDP) programs created under the American Recovery and Reinvestment Act (ARRA) of 2009. Through a quantitative analysis of customer profile and distribution circuit data collected by the Department of Energy (DOE) specific to the progress of implementation of federally funded smart grid projects, the authors were able to observe trends in the impacts of utility size and type of technology on status of completion of Electric Distribution Systems (EDS) modernization specifically. Using these data, the authors concluded that SCADA technology tended to be implemented more quickly than DA devices, regardless of utility size. In the future, this may have an impact on which technologies developers decide to use in upgrading electricity grids.

There are a variety of smart grid technologies that improve communication between the supply and demand sides of the grid, increase the efficiency of electricity transmission, reduce consumption, and increase reliability. Ghosh and colleagues at the Advanced Research Institute at Virginia Tech briefly explored the five types of projects that the 99 recipients of $3.5 billion in grants from the DOE through the SGIG program fell under. These projects improved Advanced Metering Infrastructure (AMI), Customer Systems (CS), Electric Distribution Systems (EDS), Electric Transmission Systems (ETS), and Equipment Manufacturing. Due to the interconnections between these operations, 39% of the projects incorporated several of them. However, a large percentage (57%) of projects were related to EDS, which focuses on the operations and communications of distribution technologies. The authors explored the different types, status of deployment, and significance of EDS projects. In 2009, $1.96 billion in federal funding was distributed to EDS projects which have been estimated by the US Energy Information Administration to impact over 34 million consumers, 30 million of which are in the residential sector and comprise 23% of America's total residential electricity consumers. The authors went on to define the parameters that they would refer to throughout the article as they evaluated the extent of implementation of various projects, which included the size of the utility as determined by the number of distribution circuits within a service area, ranging from very small (under 50 substations) to large (more than 500 substations). They considered two types of EDS technologies, DA devices and SCADA systems. DA devices include technologies such as automated feeder switches, automated capacitors, automated regulators, fault current limiters, smart relays, remote fault indicators, and monitoring systems; while SCADA systems are implemented in large, spread out systems where uniformity of operations and extensive communication efforts are critical. Ultimately, they examined the ratio between the number of substations that have the new technologies to the total number within each area. The ratio of substations integrating EDS technologies to baseline substations was much higher in those with SCADA systems than those with DA devices, and they were more completely implemented. The results for DA devices varied between utility sizes and appeared to be dependent on which technology was used in each specific situation. Overall, most projects that were partially funded by federal grant money through the SGIG program were near completion if they had used SCADA, and still progressing if they had used DA devices.

This quantitative case study will be useful as a model for systems progress as new technologies are implemented, particularly with respect to EDS projects. While this study focuses on the EDS side of the SGIG grant, there is much more to explore regarding the successes and failures of the other types of Smart Grid projects. Additionally, comparative quantitative studies offer the federal government a tool to evaluate the effectiveness of its investments and the results may direct investments in the future.

Smart Energy Monitor Installation results in Short-Term Energy-Use Reductions and shapes Long-Term Perceptions of Normal Use Levels

As energy demand continues to increase, more utilities are turning to smart grids to manage larger systems and the increasingly diverse array of energy sources. The first technology implemented in a system is typically the smart meter because it allows for greater communication between utilities and consumers. Giving consumers access to usage information allows them to play a role in demand-side energy management. Smart meters have been installed in many systems around the world with the intention of increasing awareness amongst consumers regarding energy use patterns, and empowering them to change their behavior in order to lower their electricity costs, and subsequently reduce greenhouse gas emissions. Studies have revealed that smart meters and other behavior monitoring devices are influential in reducing costs and energy usage immediately after installment, but few studies have examined their impacts over an extended period of time. Hargreaves et al. (2013) set out to examine the changes in consumer behavior that came from the installation of Smart Energy Monitors (SEMs) in home in the United Kingdom (UK) and to see whether the effects stuck a year later. They did this through qualitative interviews of 11 households involved in a Visible Energy Trial which revealed that initial reductions flattened out over time as the SEMs served as a monitor of normal usage rather than a tool to further reduce energy demand. These results are indicative of the challenges that utilities face in maintaining the effectiveness of technologies like SEMs as tools for energy reduction and the need for increased policy and market-driven incentives to encourage energy conservation amongst consumers.

Many SEM trials that have been carried out throughout the UK produced inconclusive results in terms of effectiveness in energy reduction in households. Hargreaves and colleagues conducted multiple surveys of a selection of households that participated in the Visible Energy Trial (VET) in England. The VET involved 275 households that were split into four test groups to observe the impacts of various SEMs between 2008 and 2010: the first group purchased the Solo, a monitor that displayed current and daily energy use and how the household was meeting or surpassing its established daily budget; the second group installed the Duet, which had additional features including boiler usage and independent monitoring of up to six appliances; the third group obtained the Trio, which contained the same features as the previous two monitors and the additional capacity to monitor the energy demand from hot water, heating, electrical circuits, and 100 appliances; the fourth group was a control group that was not able to monitor the information that was gathered through the devices installed in their homes. Quantitative data were collected regarding energy usage and cost reductions for other studies. Hargreaves et al. chose 15 households that participated in the VET, four from each technology group and three from the control group, and conducted qualitative interviews in 2009 and a year later in 2010. The follow-up interviews lasted for under an hour and were performed via telephone for 11 of the 12 test subjects who were actively using the various SEMs.

The authors focused on four areas during the interviews: monitor utilization; changes in energy education or consumption habits due to information displayed by the monitors; changes in monitor function within the household over the 12 month period; and concerns and suggestions for monitor improvements in the future. The results were generally similar across households and test groups. While three subjects had stopped using the SEMs prior to the second round of interviews for various technical and other reasons, the remaining households observed an initial peak in observation of the monitors referred to as the 'honeymoon period,' followed by a gradual decline in attentiveness. In general, users were less annoyed by the devices as they began to play more passive roles in the households in the 12 months that had passed, likely due to the repetitiveness of information displayed. Users observed that the monitors played less of a role after the initial learning period as they gained a grasp of their energy use patterns, and this sometimes resulted in the movement of the monitors and a transformation in how they were used in the households. Some participants mentioned that they found different information from the monitors to be valuable over time, focusing more on specifics than total consumption. They also established the idea of a normal use level that they would try to maintain rather than continuing to reduce energy demand. Many households felt that their energy awareness and practices had changed due to the presence of the SEMs. By visually presenting data, the monitors provided an easily interpretable depiction of the effects of various behaviors on energy demand. Consumers focused on their initial ability to identify areas for significant reductions, but over time this disappeared as they settled into their normal use levels at which point they would be compromising convenience and well-being in order to conserve energy. The authors found that most interviewees were not particularly driven to reduce energy consumption by environmental incentives. Trial participants were also observed to face many of the same challenges that they foresaw in the first round of interviews. Amongst these limitations were difficulty in modifying routines and social dynamics within households. Hargreaves et al. identifies three main obstacles that grew over the 12 months between interviews; household disagreements, inflexible demand, and political and market limitations.

The authors concluded that SEMs led to initial changes in household energy consumption in the short-term but reinforced the concept of a normal level of demand in the long-run. Interviewees indicated that the monitors allowed them to observe the level of consumption that was typical for them and worked to maintain that level rather than continuing to identify areas for conservation. In addition to complex household relationships, Hargreaves et al. identify the role of public policy and market conditions in reinforcing these norms, particularly focusing on three issues. First, there was a clear and unpredictable difference in how each household used the monitors which should be addressed in future demand-side management approaches with smart grid technologies. Second, the authors noticed that users were not likely to reduce their consumption without external incentives and support. Lastly, SEMs did not prompt householders to question their concepts of normal use which will provide a challenge in energy conservation efforts as societal energy demands continue to increase. The authors suggested that policies be used to limit energy use of individuals in order to challenge the concept of normal use and further change energy consumption patterns such that SEMs would be imperative to assist users in complying with energy limits.

Non-cooperative and Cooperative Demand-Side Smart Grid Day-Ahead Optimization Practices Result in Reduced Reliance on Peaking Power Plants

Electricity production and usage are closely related with the impacts of climate change because many forms of energy come from the combustion of fossil fuels and emit high amounts of carbon dioxide into the atmosphere. Smart grids, electricity networks that incorporate communication devices and allow energy to flow both to and from consumers and producers in order to increase efficient energy usage, offer a promising solution to many problems that are associated with traditional energy grids. By accommodating individual energy producers and storers, in addition to traditional production methods, smart grids provide a necessary modernization of energy management systems. Atzeni et al. (2013) studied several optimization methods that cater to different types of users on the demand side of smart grids. Through day-ahead demand-side management, the authors monitored the behavior of noncooperative and cooperative energy users in order to determine which was more beneficial in reducing costs among consumers. Both simulations demonstrated that active electricity users had the potential to reduce costs by distributing generation and storage according to time-slot dependent rates, regardless of whether they were strategizing individually or within a group, thereby stabilizing the load throughout the day and improving predictability of aggregate demand.

Recent developments in smart grid technology and management have focused on the demand side of energy and consumer behavior. Energy supply is reactive to consumer demand, so changes to the system have involved educating users and providing incentives to use less energy in a smarter way. Italo Atzeni and his fellow authors approach this topic from the demand-side as well by applying game theory to cooperative and non-cooperative day-ahead energy consumption scheduling. They gathered hour-by-hour energy consumption data for a 24 hour period from a variety of consumers and used the information to produce algorithms to represent usage. Both passive and active users were accounted for in the algorithms; passive users being those that simply accept electricity from the grid and active users being those who possess energy storage or production capacity. Active users maintain the ability to function independently from the grid at times of peak usage by tapping into their own stored or personally produced sources and can take advantage of non-peak energy prices by storing energy to use in the future. The authors constructed two different models in which active users could determine their electricity usage for the following day with the intention of reducing personal energy costs by consulting a pricing model. The first model incorporated non-cooperative users in which users considered only their own needs. The second model was cooperative and allowed groups of users to collaborate in deciding how much energy they would use and during what time slots. Factors accounted for in the models included the type of users, non-dispatchable and distributed generation power, the storage capacities and characteristics of energy storage devices, and the varying cost per unit of energy.

Each model resulted in similar load shifts from peak hours to non-peak hours, thus evening out the aggregate electricity demand curve. Consequently, both active and passive users observed a decrease in their electricity costs. While each method produced the same result, the authors concluded that the cooperative optimization practices are probably superior to the non-cooperative ones because they offer optimization on a larger scale which contributes to greater predictability and stability within the system. The authors also concluded that users who implemented their own energy production and/or storage technologies experienced the highest energy savings. Ultimately, the findings from these models demonstrated the ability that consumers have when using smart grids to stabilize electricity usage such that high-emission power plants that are only used in times of peak power are no longer required to meet society's electricity demand.

As the global population continues to grow and individual electricity usage increases, reforming the energy grid has become a regulatory priority in an effort to confront climate change. Addressing energy consumption from the demand-side requires educating consumers and increasing communication regarding rates and usage trends, but it does not necessarily require extensive new infrastructure like supply-side advancements demand. As the authors demonstrated, there is much that can be done by consumers to improve energy efficiency and reduce carbon emissions. The continued expansion of renewable energy sources and increasing prevalence of energy storage vehicles and devices that are present on the electricity grid will allow smart users to rely on clean energy and to store excess when it is available in order to reduce their reliance on fossil fuels.

Maximizing the Storage Potential of Electric Vehicles in Smart Grids that are increasingly dependent on Renewable Energy Sources

Large scale renewable energy projects, as well as small installments on houses and businesses, have the ability to transform the composition of energy sources that fuel society. While such sources are beneficial in reducing greenhouse gas emissions that result from power production, they provide new obstacles for the use and storage of energy. As renewable energy sources continue to produce larger quantities of energy, the electricity grid will have to adapt to this decentralized and unpredictable production. Ridder et al. (2013) examined the feasibility of charging electric vehicles (EVs) in a smart grid scenario and their ability to compensate for shortages and surpluses that occur in the energy market due to unpredictable energy supply. Utilizing behavioral information about the EV users, EV capacity, and the capacities of charging stations, the authors created a model to simulate EV potential in Flanders, Belgium. The coordination algorithm that was determined to fit the population resulted in the proper distribution of EVs amongst available charging stations so that capacity was maximized. In the second scenario, the authors demonstrated the ability of EV users to take advantage of varying prices of energy by strategically charging when excess day-ahead energy was available and selling energy back to the grid when demand exhausted supply and only day-of, or imbalance market energy was available. Ultimately, the study demonstrated that as imbalance market prices increased compared to day-ahead prices, flexible chargers were able to save money and charger/dischargers were able to make money by discharging energy at high market prices.

In order to determine the role of EVs in conjunction with variable tariffs in matching energy consumption to supply, Fjo D. Ridder and colleagues utilized algorithms and activity based models to simulate the behavior of the population of Flanders, Belgium. The scheduling algorithm utilized in this study takes into consideration EV owner behavior, capacity and characteristics of current EV batteries, and availability of smart grid technologies. The FEATHERS system, an activity based model specific to Flanders, created a daily schedule for each subject within the simulation. The system created data for the 6 million people in the artificial population that reflected the census data for the area, incorporated land-use data and applied it to 2368 traffic analysis zones (TAZ), combinations of times and distances for different routes based on type of transport, and decision-making tendencies of individuals based on their corresponding socio-economic data. The FEATHERS model was able to mirror the decision-making processes of the population, and thus could be used to create day-ahead demand schedules. Using data that represented the city of Flanders, the authors constructed a simulation under the assumptions that there would be 200 EVs spread out over 56 parking areas, equal electricity tariffs determined by the Belgian Power Exchange, and equal maximum charging powers, battery capacities, initial charges, average consumption while driving, and driving speeds for all EVs. Finally, a utility function was attributed to each EV to calculate the charging cost and adjust for lost power. Many of the same assumptions used in the activity-based model were applied to the utility functions.

When the first application was performed, multiple iterations resulted in a balancing of the distribution of EVs charging at each location based on the total amount of power available daily. This demonstrated that EV owners would quickly adjust to the supply of power at charging stations and adjust their locations accordingly. The second application sought to minimize costs for electricity retailers. In this simulation, retailers purchased power using day-ahead models to predict the demand. Since electricity prices that consumers paid fluctuated throughout the day based on demand, the prices would be lowest when there was less actual demand than the retailer had anticipated. Another scenario could occur if the retailer underestimated the electricity demand and some had to be purchased that day. If there were a plethora of power available, these prices would be more competitive than day-ahead prices. The opposite scenarios were also possible and would result in higher prices.

The authors simulate three consumption situations in the second application; the no-strategy benchmark users, charging only users, and users with the ability to charge and discharge. The results demonstrated that EV users were sensitive to price and would adjust their usage in order to minimize costs. In the second case, users determined when they charged their vehicles and they saved 60%, on average. In the final case, users were able to charge when the price was the lowest as well as discharge when the price was the highest. On average, this scenario resulted in profits of 128 Euro for each user, each year.

While there are not enough EVs in use currently to play a large role in the electricity grid, they do have the potential to bridge the gap between the shortages of supply or demand of electricity which are often brought about suddenly by renewable energy sources. Charging vehicles when surplus energy is available prevents it from going to waste by providing a form of storage, and discharging power during shortages reduces the need to suddenly generate more electricity. And by giving the user control over the demand schedule, EVs do not create the privacy issues that other demand side management methods do by relying on an information processing controller. By providing a consumer-driven method for storing electricity, EVs have the potential to flatten the price curve for electricity throughout the day and more evenly distribute demand in conjunction with supply.

Advanced Information System Technologies in Smart Grids Allow for Increased Utility Effectiveness in Improving Energy Efficiency

As climate change progresses, concerned parties have turned to the electricity grid as a critical target for making large improvements in efficiency and reductions in greenhouse gas emissions. Engineers are scrambling to construct a more efficient electricity production and distribution system as the global population continues to increase and society becomes increasingly reliant on energy to fulfill its needs; this new system is referred to as the smart grid. As a hybrid between smart grid and demand-side management program efficiency strategies, information systems provide a link between the demand and supply sides of the grid and are being implemented with varying degrees of success throughout the United States. While a variety of smart grid technologies exist, Corbett (2013) evaluated the efficiency improvement abilities of three specific types of information systems; AMR (automatic meter readers) that provide the utility companies with usage information from customers, smart metering that allows for the transmission of information back and forth between the utility company and the consumers, and net metering which involves two-way communication and allows the consumer to sell power back to the grid. Corbett explained the significant potential of information systems in making energy distribution more sustainable, then proceeded to outline her two hypotheses: the first being that new smart grid technologies such as advanced metering infrastructure (AMI) would improve the information processing abilities of utilities and consequently their effectiveness in demand-side management; and the second countering the first by positing that extensive information processing improvements in utilities could reduce efficiency in demand-side management due to an inability to properly manage additional quantities of data. Using a hierarchical regression to evaluate data collected from electric utilities across the nation, the author surprisingly discovered that both hypotheses, although conflicting, proved accurate. While there was significant consensus that demand-side management information technology programs improved energy efficiency, the data also pointed to the fact that increased data collection reduced effectiveness of utilities when utilized improperly.

While most research evaluating smart grid technology and demand side management program implementation focuses on the consumers of energy, Jacqueline Corbett examined the role of utility companies and the degrees to which they have successfully utilized smart grid technologies to increase energy efficiency by gathering, interpreting, and reacting to usage information of customers. Corbett began by analyzing the organizational information processing theory (OIPT) to derive the significance of new technologies in the performance of utility companies. OIPT suggests that a company's effectiveness is shaped by its ability to connect its processing requirements to its capacity, or, in other words, its ability to accommodate and react to its informational demand.

The hypotheses were tested against data submitted by utilities serving residential consumers throughout the US in the "Annual Electric Power Industry Report" from 2009. Corbett observed the impact of a control group and three types of AMI technology, automated meter reading devices (AMR), smart meters, and net meters, on effectiveness of demand-side management programs. The author controlled for variation between the data due to size of utilities, location, season, type of ownership, and other factors that could have impacted effectiveness besides the technologies themselves. The model concluded that 73.2% of the variation in effectiveness between utilities was attributable to the demand-side management programs. The results for the AMR technology showed a negative correlation, providing support for the second hypothesis, meaning that implementation of automated meter readers actually had a negative impact on effectiveness in demand side management efficiency efforts. The smart meters and net meters supported the first hypothesis and confirmed that the more advanced technologies were generally successful in improving utility effectiveness.

Corbett's exploration of demand-side management efforts in smart grids from the point of view of utility companies and their effectiveness in implementing efficiency improving methods demonstrated the ability of information systems technologies to have an impact on effectiveness, though the nature of the impact varies depending on the technology. While more advanced technologies were observed to have predominantly positive impacts, the one-way communication provided by AMR metering devices had a negative impact on effectiveness. The author speculated that this was a result of the utilities' inabilities to properly process the additional data they collected. While smart meter technologies are intended to provide information that will help both utilities and consumers, that cannot occur unless processes are in place to interpret and apply the data. Thus, less advanced technologies were believed not to support this efficiency enhancing process, and in fact, resulted in the opposite effect. The author suggests that future studies observe the impacts of continually improving technologies on effectiveness and utilize data from more progressive countries, such as many European countries, where more smart grid technology is implemented.

Maximizing Flexible Electricity Use and Minimizing User Inconvenience through Load Balancing of Smart Grids

The electricity supply has traditionally been dictated by consumers. Consumers demand varying amounts of energy depending on their instantaneous needs and suppliers are left to use whatever resources are necessary to meet their demands. As populations grow and electricity demands per capita increase, the discrepancy between demand and sustainable supply levels continues to widen. The smart grid may have the potential to mediate the conflicting objectives of consumers, who prefer supply levels that correspond with high levels of convenience for them according to their preferences, and suppliers, who would benefit from producing at a more constant rate. Hassan et al. (2013) explore the plausibility of load balancing, which has been enabled by smart grid technologies, as a method of balancing demand to more closely align with reasonable supply levels. Electricity load balancing enables consumers to designate essential devices, meaning that they provide services that users cannot live comfortably without, and non-essential devices, which have more flexible usage requirements. Energy suppliers can take advantage of non-essential devices that consumers declare are flexible and provide the necessary electricity to run them when it is most convenient given the current supply. The authors constructed two models, the Grid Convenient (GC) and User Convenient (UC) Schedules, which served as the two extremes for how loads can be distributed over time. They proceeded to evaluate the degrees of deviation from grid objectives and the impacts on inconvenience levels imposed on consumers for a variety of levels of flexible devices within the total load. Hassan et al. concluded that increasing the number of flexible loads in the smart grid fulfills producer objectives and only minimally increases user inconvenience to a certain flexible load level, beyond which inconvenience levels continue to increase significantly and utility gains only improve mildly. The authors hoped to shed light on the benefits and potential of load balancing to more predictably and sustainably utilize and manage the smart grid so utility companies can create incentives that will encourage consumers to expand their electricity use flexibility in the future.

While prior studies have explored the potential energy supply savings from load balancing, Naveed Hassan and colleagues were determined to consider these benefits in conjunction with the inconvenience levels subsequently imposed on consumers by exerting more control over their electricity use patterns. The authors began by defining the variables, which included: essential and flexible load types, which were based on the necessity of timely usage of devices, levels of inconvenience for users based on deviation of loads from their preferred time slots, and the GC and UC schedules. The GC Schedule grouped together all essential and flexible loads and distributed them evenly to construct a flat electricity demand curve which was optimal for energy suppliers. The UC Schedule, on the other hand, allowed consumers to distribute their loads according to their preferences and thus represented the current day energy consumption distribution. The schedules served as the extremes in this study, and the authors sought to discover the point at which there were enough flexible loads to flatten the demand curve without significantly inconveniencing consumers, which was somewhere between the models. They did this by creating new load combinations under the assumptions that only flexible loads could be shifted and changing time slots would increase inconvenience levels. The simulation consisted of one hour time slots, baseline consumption household appliances, loads that varied between one and five kilowatt hours (kwh), and 100 total appliances that required between one and five time slots each. The authors created a sub-optimal algorithm using Multi-Processor Scheduling to assign loads to time slots. They simplified this algorithm so they could manipulate the variables more easily by assuming that all flexible devices had equal power consumption, varying time slot requirements, and deeming task order irrelevant. Hassan and colleagues proceeded to run trials of the algorithm with varying numbers of flexible devices and measured the deviation from the GC and UC schedules, as well as the corresponding inconvenience levels.

Trials with varying numbers of flexible devices within the 100 device load revealed that higher percentages of flexible devices did result in the flattening of the electricity demand curve. Subsequently, however, there were significant increases in inconvenience levels as the share of flexible devices increased, because consumers had less control over electricity usage for those appliances and their preferred time slots were not always available. Between 40 and 100 flexible devices, the authors noticed that there was only a slight 5.1% improvement in terms of load balancing, while there was a significant 52.7% increase in inconvenience levels. The authors concluded that increasing the share of flexible devices in the load profile was beneficial to the utilities and relatively harmless to customers until a certain level, at which point the load balancing began to cost more per unit in terms of increases in consumer inconvenience. Thus, Hassan and colleagues recommended incentivizing a relatively small number of flexible devices so that producer benefits were observed but inconvenience to customers was minimized.

Examination of Smart Grid Implementation in Six European Countries Reveals Potential Carbon Emission Reductions with Policy and Technological Support

Carbon emissions are greenhouse gases and are often targeted for reductions in order to slow the progression of climate change. The energy sector, in particular, is seen as an area with significant potential for minimizing emissions since it is responsible for such a high percentage of society's atmospheric carbon contribution. In Europe, a plethora of smart grid technologies has been installed and more are being designed in order to increase efficiency of electricity production and transmission. Darby et al. (2013) examined six national energy markets in the European Union (EU) in order to determine how carbon emission reductions occurred with the implementation of technologies and policies, market characteristics that were conducive to reductions, the areas with the greatest potential of achieving emission reductions, and the areas in which the new systems would be most effective. They collected a variety of quantitative and qualitative data from the German, Austrian, French, Spanish, Portuguese, and British markets in order to predict the emission reductions produced under three conditions: no smart grid implementation, smart grid technology implemented without legal or economic support for users, and smart grid technologies installed and supportive legislation and market conditions adopted. The models revealed that reducing demand and increasing energy efficiency would result in considerable emission reductions due to the composition of energy sources. The authors concluded that the most potential for lowering emissions existed within the residential sector and policy changes and market incentives were needed in order to promote action on the part of the consumers in order to provide the required demand reductions. The results were similar across the six nations, for the most part, and slight variations were attributed to regulatory differences.

The EU and various nations around the world have set challenging goals for reducing greenhouse gas emissions, partly through increases in renewable energy sources. The authors identified three ways in which energy production and distribution technology developments have been shown to lower carbon emissions in the energy industry: through a decrease in demand, typically accompanied by increases in efficiency; through supplier manipulation of demand at peak times through price structures or flexible use agreements; and increased availability of renewable energy sources. Darby and colleagues performed the methodologies to measure the potential of smart grids for greenhouse gas reductions (SG4-GHG) study in accordance with the Digital Agenda for Europe's European Commission Taskforce on Smart Grids to help determine the future role of smart grids in the EU. Two models, the Zephyr pan-European market model and the ancillary services model, were constructed to determine the impacts of various quantitative and qualitative functions of the smart grid, on the demand-side, supply-side, and through the interconnection of systems, on carbon emission reductions from smart grid implementation. These functionalities revealed that there were two primary sources of reductions; reductions in the volume of energy produced and decreases in the emissions per quantity of energy generated.

The authors proceeded to outline metrics that characterized the markets they examined and envisioned. In terms of demand-side driven reductions, Darby and colleagues identified four methods that consumers were able to use to modify their energy demand as enabled by smart grid technologies; overall reductions, static peak reduction, dynamic peak reduction, and continuous balancing. Additionally, applicable United States (US) and Australian data were used to fill in for the information void in EU data pertaining to multiple desired metrics. The first model used the respective national data collected for overall and peak reductions and the second model displayed the potential impacts of continuous balancing. Each model was run for three variations of 2020 scenarios: baseline, in which there is no smart grid; expected, in which there are new smart grid technologies but limited to no legislative support; and feasible, where technology and legislation continues to adopt and expand. The models required various assumptions be made, and the authors did so through extensive research in the field to produce the most accurate predictions possible.

The results of the models indicated the potential for reductions in each of the six countries examined in the study. Germany and Great Britain had the highest potential and the authors attributed this to their use of coal, a carbon-intensive energy source, in their current energy profiles. It was also noted that opportunities for savings were higher within the wholesale energy markets, as opposed to the ancillary markets. The most significant factor in emissions reductions for all countries was the lowering of carbon intensity through renewable sources and greater efficiency measures. Lower total demand was also noted as a significant factor. Ancillary markets, which store power and release it as needed, saw smaller reductions because even as renewables provided a higher share of power at certain times of the day, fossil fuels were still relied upon for the majority of the time. Overall, Darby and colleagues were confident that reductions in demand had more influential impacts on emissions than peak shifting efforts. Notably, small reductions in demand corresponded with large reductions in emissions because they reduced reliance on fossil fuel production methods. Market connections were also identified to be useful in several of the countries, particularly those producing excess renewable energy that could be sold to neighbors. Thus, the authors concluded that smart grid programs should focus on demand reduction and efficiency improvements. This will require increases in policy support for customers and market incentives for them to change their behavior in order to fully take advantage of the newest grid technologies. Darby and colleagues recommended that further research focus on the long-term impacts of smart grid technologies, grid interconnection expansion benefits, and the implementation of more pilot programs. Ultimately, smart grid technologies resulted in significant carbon emission reductions under the favorable technological and political conditions in the EU.

Using Policy Approaches to Reduce Carbon Dioxide Emissions from the Electric Grid through Demand and Supply-Side Impacts

While energy production is widely acknowledged as a significant contributor to climate change, there is a discrepancy in opinion about what the most effective solution is to cut back on emissions. The most commonly addressed method of bringing about a smart grid is through new technologies that have the potential to improve distribution efficiency, encourage demand side management behaviors, and reduce the emissions associated with the production process. Policy change, however, is another route that has the potential to be more efficient in reducing emissions in the short term as technological developments are in progress. Li et al. (2013) examined the potential of several types of policy initiatives to modify electricity operator behavior in order to reduce CO_2 emissions while continuing to meet energy demand. Basing their assumptions on the energy profile of Michigan, the authors created three models to represent different policy approaches: the first served as a baseline and represented the present energy cost and load distribution, the second imposed demand-side financial penalties for CO_2 emissions, and the third created a carbon disincentive that produced a new pricing scheme for energy sources in terms of emissions. The models revealed that the carbon disincentive approach was the most successful in changing operator behavior because it increased the production costs associated with selling dirtier energy without raising prices considerably for consumers. The roles of plug-in electric vehicles (PEVs) and renewable wind energy were also examined, separately and together, and Li and colleagues concluded that the technologies were most beneficial when implemented jointly so that the low-cost, emission-free wind energy would supply the majority of the energy demand increase brought about by the PEVs, while the PEVs would serve as storage for wind energy produced in periods of low demand. Ultimately, the authors argued that a carbon disincentive program is the preferable policy measure to address carbon emissions because it is not a tax placed directly on carbon, but rather it restructures the costs of energy sources in order to incentivize suppliers to use cleaner sources first, even if it is more costly to do so.

Without external forces, energy suppliers are motivated by the cost of energy in deciding which sources to use to comprise the load that is demanded by their customers. Currently, nuclear power is the cheapest form of energy and does not result in measurable carbon emissions. As a result, nuclear power has been chosen by operators and approved of by environmentalists as the first power source, when available, to satisfy demand. In many areas, such as Michigan, the next most prevalent energy source is coal. Coal-fired power plants are among the heaviest polluting in terms of CO_2 emissions per output of energy, but because coal is cheaper than alternative sources, it is used frequently. Natural gas is another common source and while slightly more expensive than coal, it has a smaller environmental impact. Other forms of renewable energy, such as wind power, are more expensive, or are unpredictable and are thus not commonly relied upon by energy suppliers. The dichotomy between price of power and CO_2 emitted poses a challenge to those who are trying to reduce the environmental impacts of the electricity grid. Li and colleagues evaluated the success of different policy measures intended to change the tradeoff between emissions and costs so operators would use lower emitting energy sources first. They began by listing the assumptions behind their models. The hypothetical grid was price-inelastic, the grid operators were assumed to utilize cheaper power sources first, costs of reserves were considered, PEV characteristics were standardized, and emission data for each type of energy production, including wind, were acquired and interpreted. The authors used this information to evaluate the costs and emissions that resulted from the baseline scenario, the carbon penalization policy, and the carbon disincentive policy.

The baseline scenario did not take carbon emissions into consideration and produced a cost-minimizing sequence of energy sources. As a result, the first 2040 megawatts (MW) of power were nuclear, the subsequent 4800 were supplied from coal, and the remaining demand was satisfied with production from natural gas. The second model enacted a demand-side penalty based on emissions. The authors experimented with different prices for each ton of CO_2 emitted, ranging from $10 to $55. While the penalty cost did alter behaviors, it created significant inconveniences for PEV owners who began charging their vehicles at different times and changing the demand composition. As a result, only moderate emission reductions were observed. The final model incorporated the supply-side targeting disincentive policy, which added between $0.05 and $20 to the generation cost incurred by power plants per ton of carbon emitted. The disincentive was thus felt by the grid-operators who revised the composition of their load based on the new prices for each type of power. The costs of renewable energy and other lower emitting sources became more affordable when compared to coal, and were thus increasingly relied upon to meet demand. Meanwhile, the cost of emissions was not felt considerably by the grid operator or consumers due to its "revenue return mechanism." When the disincentive was $0.05 per ton of CO_2, emissions were 0.13 percent lower and consumer cost increased 0.04 percent. When the disincentive was increased to $20 per ton, emission reductions increased to nearly 25 percent and cost increases almost reached 20 percent. More heavily polluting power plants were negatively impacted by the disincentive and experienced lower profits, while renewable energy producers saw higher profits.

The results of the models indicated that the disincentive policy was the most influential in reducing carbon emissions. By altering the composition of the power supply that utility operators used to meet consumer demand, this policy encouraged cost-driven suppliers to consult low-emitting sources before relying on higher polluting methods, such as coal-fired power plants. This method, as compared to the current scenario and the direct penalty option, altered behavior on the supply and demand sides and was significantly more effective in reducing CO_2 emissions. The models also revealed the ability of PEV and wind power implementation to reduce emissions, and the authors concluded that the technologies should be implemented together in order to improve effectiveness. While the impacts of a carbon disincentive as applied in this study would render different results depending on the composition of energy sources in a geographical area, this policy approach could have significant impacts in the transition to smarter electricity grids.

Conclusions

Smart grid developments have proven to have varying degrees of success depending on their abilities to alter consumer and producer behaviors. Studies have shown that users and producers of electricity are primarily driven to modify their energy consumption and purchasing patterns through cost reductions. As a result, policy incentives will be required to compliment technological advancements in order to bring about significant emission reductions. Communication technologies, for example, allow consumers to monitor their usage while producers can create price variations depending on demand and supply conditions that will drive consumption patterns. Demand-side management measures are crucial in involving energy consumers and allowing them to play a more active role in the energy sector through engagement with the grid. Technological advancements also allow for the integration of renewable energy sources and storage devices, such as electric vehicles. Without policy and market incentives, however, behavioral changes that take full advantage of these technologies will not be possible or worthwhile for the demand or supply sides of energy. With continued investment of both public and private funds, smart grid developments will not only strengthen existing electricity grids to protect them from the potentially harmful impacts of climate change, but they will also allow the energy sector to reduce its greenhouse gas contributions that perpetuate these potential threats.

References Cited

Arnold, G., 2011. Challenges and opportunities in smart grid: a position article. Proceedings of the IEEE, 922–927.

Atzeni, I., Ordonez L., Scutari, G., Palomar, D., Fonollosa J., 2013. Noncooperative and cooperative optimization of distributed energy generation and storage in the demand-side of the smart grid. IEEE Transactions on Signal Processing 61, 2454–2472.

Crobett, J., 2013. Using information systems to improve energy efficiency: Do smart meters make a difference? Information Systems Frontiers 10.1007/s10796-013-9414-0, 1–14.

Darby, S., Stromback, J., Wilks, M., 2013. Potential carbon impacts of smart grid development in six European countries. Energy Efficiency, 725–739.

Ghosh, S., Pipattanasomporn, M., Rahman, S., 2013. Technology deployment status of U.S. Smart grid projects—electric distribution systems. IEEE Innovative Smart Grid Technologies, 1–8.

Guccione, L., 2013. The micro(grid) solution to the macro challenge of climate change. Rocky Mountain Institute Blog. http://blog.rmi.org/blog_2013_10_02_microgrid_solution_to_macro_challenge_of_climate_change.

Hargreaves, T., Nye M., Burgess J., 2013. Keeping energy visible? Exploring how householders interact with feedback from smart energy monitors in the longer term. Energy Policy 52, 126–134.

Hassan, N., Wang, X., Huang S., Yuen, C., 2013. Demand shaping to achieve steady electricity consumption with load balancing in a smart grid. Innovative Smart Grid Technologies (ISGT), 1–6.

IPCC, 2013. Summary for Policymakers. In: Climate Change 2013: The Physical Science Basis. Contribution of Working Group I to the Fifth Assessment Report of the Intergovernmental Panel on Climate Change [Stocker, T.F., D. Qin, G.-K. Plattner, M. Tignor, S. K. Allen, J. Boschung, A. Nauels, Y. Xia, V. Bex and P.M. Midgley (eds.)]. Cambridge University Press, Cambridge, United Kingdom and New York, NY, USA.

Li, C., Peng, H., Sun, J., 2013. Reducing CO_2 emissions on the electric grid through a carbon disincentive policy. Energy Policy, 793–802.

Lyster, R., Byrne, R., 2013. Climate change adaptation and electricity infrastructure. Sydney Law School Research Paper No. 13/23, 1–41.

Ridder, F., D'Hulst, R., Knapen, L., Janssens, D., 2013. Applying an activity based model to explore the potential of electrical vehicles in the smart grid. Procedia Computer Science, 847–853.

U.S. Department of Energy's Office of Policy and International Affairs, National Renewable Energy Laboratory, 2013. U.S. Energy Sector Vulnerabilities to Climate Change and Extreme Weather [Zamuda, C., Mignone, B., Bilello, D., Hallett, K., Lee, C., Macknick, J., Newmark, R., Steinberg, D. (eds.)]. http://energy.gov/sites/prod/files/2013/07/f2/20130716-Energy%20Sector%20Vulnerabilities%20Report.pdf.

U.S. Department of Energy Office of Electricity Delivery & Energy Reliability. Smart Grid. http://energy.gov/oe/technology-development/smart-grid.

U.S. Department of Energy Office of Electricity Delivery & Energy Reliability. Smart Grid Investment Grand Program – Progress Report (July 2012). http://energy.gov/oe/downloads/smart-grid-investment-grant-program-progress-report-july-2012.

4. Exotic Biofuel Technologies

Chieh-Hsin Chen

Form the industrial revolution to the present, we have used more than half of the world's fossil fuel. The fossil fuel produced over millions of years could be depleted within a hundred years. Energy sources have become some of the greatest issues facing the human race. Scientists have been working on alternative sources such as nuclear power plants and solar panels; however, energy depletion is not the only problem. Global warming caused by the emission of greenhouse gas, especially CO_2, is also one of the challenges. Some recyclable energy sources such as solar panels and hydropower plants are not adequate to substitute completely for fossil fuels, scientists thus have been working on development of biofuel, a fuel source based on living matter. Although biofuel is a young researching field, there are multiple forms of it and it has the potential to substitute for fossil fuel. Biofuel still has many shortcomings but it could possibly become the next primary source of energy for cars and electricity.

Most common types of biofuel utilize glucose or other carbon compound from plants to form methanol or ethanol as a combustion source for energy. This type of biofuel does not significantly reduce in the emission of greenhouse gas, but the raw fuel source, plants or algae recycle the carbon dioxide into usable energy through photosynthesis. This method creates a cycle of carbon in the ecosystem through the process of producing energy; furthermore, this common method is very efficient, almost as efficient as thermal power plants. The biofuel can become alternative to fossil fuels, solving the issues of excessive atmospheric carbon. Liquid biofuel may also be used in vehicles substituting for gasoline. In fact, biofuel has been mixed with gasoline for many years and there has been a gradual increase in the use of biodiesel.

There are also multiple methods and types of biofuel in the field. Biofuel cells are very interesting developments. Although they have lower efficiency and have multiple shortcomings that needed to be improved, they utilize carbon compounds straight from the fuel source to produce electricity. The mechanism of biofuel cells uses enzymes to break down the fuel sources while creating electro-potential energy. Recent study of biofuel cells has looked into various aspects of modifying: the cathode and anode, fuel sources, and low cell life. Some of the greatest disadvantages of biofuel cells are the low efficiency and the cell life. To increase a biofuel cell life, scientist had tried various methods to keep enzymes from denaturing, and so far there has been some positive progress.

To improve the development of any type of biofuel, evaluations are important; and there were also numerous evaluations on different aspects of biofuel production. The assessment of possible fuel sources is one of the major factors affecting the progress of development. Before mass production, scientists work on evaluating the variables of biofuels made from different sources. Some aspects of the variables are based on environmental perspectives and the efficiency of the biofuel. Understanding the source and production of different biofuels allows scientists to eliminate the less ecofriendly or less efficient biofuel from further consideration.

After all the research and evaluation, the government needs to provide an effective strategy for mass production and utilization of biofuel. In some developing countries, they adopt biofuel substitution plans to possibly make biofuel their primary energy source in the future. Scientists also evaluate the government plans and give suggestion on the progress of the strategy.

Tracking Reusable Biofuel From Crops: Carbon and Solar Energy

Technology for production of biofuels has been a popular research area for many biochemists in the past ten years because finding alternatives to the use of fossil fuels can have important effects on the future of human life. Although the technology is not able to produce low cost, efficient biofuels yet, Borak *et al.* (2013) promote the possibilities and efficiency of biofuel production through crops. Their "PETRO approach" is used to evaluate new crops, not only on the capture of solar energy but also the capture of carbon in atmosphere. The reverse of the carbon combustion cycle naturally occurs within plants, which use photosynthesis to convert carbon dioxide from the air to usable fuels. The energy source products from various crops are similar, but the efficiency of conversion of sunlight into energy varies. The authors summed up the total loss of energy during each process and organized data on the final energy content for each crop. They showed that using the PETRO approach for evaluating potential crops as biofuels can lead to more detail-based discussion in the scientific community.

Because of the depletion of fossil fuels and the increase of atmospheric carbon dioxide from combusting fossil fuels, scientists are eager to find alternative fuels. The two main barriers to production of alternative fuels are the costs and shortage of potential stocks; thus the production of liquid fuels from crops has become one of the top environmental goals for future research. On a fundamental level, the concept of biofuel is replacing the process of mining with the process of agriculture; the process shift is significant because biomass has significantly higher carbon oxidation states than fossil fuels. The use of biofuel essentially reverses the combustion of carbon-based fuels, capturing the byproduct CO_2 converting and storing as usable energy via carbon fixation utilizing plants.

To further enhance biofuel production, a detailed evaluation of efficiency of the crops and productivity of conversion of energy will be useful; however a systematic methodology for evaluation is currently lacking. Different research groups studying in different geographical areas use a variety of non-comparative assumptions and approaches to calculating yields. Data consistency became one of the difficulties for further biofuel research. The authors introduced the Plants Engineered To Replace Oil (PETRO) approach that included the input of raw materials (sunlight, carbon dioxide and water), trace process of conversion by plants, and the output of liquid fuels. Photosynthesis reverses the combustion of fuels and stores the carbon energy along with solar energy in the plant; the production of biofuel extracts and concentrates carbon energy from plants, usually as ethanol, converting it into usable fuel.

Although plants have evolved effective photosynthetic pathway to capture light, they are not as efficient as we would like; the result of evolution does not maximize the conversion of sunlight for producing food or fuels. That C3 plants only utilize 4.6% of the solar energy and C4 plants utilize 6.0% suggests substantial room for improvement, even with these low capture levels of photons much of the captured energy is subsequently lost.

The authors looked into the difference in loss of carbon energy between C_3 and C_4 plants in four levels: captured, harvested, purified, and processed. C_3 plants lose half of their usable carbon energy in photorespiration and more in respiration. C_4 plants lose less in respiration, but C_4 plants lose more than half seasonally. Overall the final usable carbon energy is about 0.69% in C_3 plants and 3.0% in C_4 plants. But there are also some losses of carbon energy in the processing steps.

Although energy loss in the processing steps is small compared to the loss from photorespiration or seasonality, it is the first step that can be improved through conventional engineering. With careful genetic selection and engineering of crops, we will be able to control the seasonality and other growth process of crops. The authors introduce data that show the differences in final fuel products of four common crops used in biofuels: maize, soybean, sugarcane, and switchgrass. The final energy content of the four different crops is very similar, but they have significant differences in overall fuel yield. Sugarcane contains the most overall fuel yield and soybean the least.

Life Cycle Assessments of Biofuel Options

The emergence of biofuel provides an alternative source of power and also tends to reduce greenhouse gas emission compared to petroleum fuel. Life cycle assessment (LCA) is used to evaluate environmental impacts from the production and use of biofuel measuring the greenhouse gas emissions versus the energy produced. The first generation of biofuel provides little greenhouse gas reduction and multiple indirect side effects. Second-generation biofuels are intended to improve on those issues. Kendall and Yuan (2013) discussed the results of LCA for three of the most researched second-generation biofuel pathways: cellulosic ethanol, algae biodiesel, and sugarcane ethanol. LCA is useful tool when applied consistently across production system to select the most environmentally preferred crop. However, LCAs of the second-generation pathways lead to inconclusive results.

Evaluation of potential crop feedstock is one of the most important steps for biofuel research. Life cycle assessment should help address this question but when the authors looked into recent studies of LCA they found high level of uncertainty because of model and method-induced variability and variability and uncertainty due to real differences in the modeled systems.

The first stage of LCA is definition of goal and scope. The attributional approach evaluates the performance of systems by simply tracking flows of input and output on a life cycle basis. A consequential approach considers changes to a system that may directly or indirectly affect the environment. One of the most significant consequential approaches of biofuel is land use change (LUC), which the feedstock may directly influence sometimes competing with food production, other preexisting land uses, and gradually causing the food market price to raise. The next stage of LCA is a life cycle inventory (LCI). The third stage is impact assessment identifying the volume and timing of emission of greenhouse gas.

The cellulosic ethanol pathway uses lignocellulosic plant matter such as corn stover and switchgrass as its feedstock. This pathway requires conversion to simple sugar and fermentation to ethanol. Corn stover has been widely researched as a possible source among crop residue, but there are still various issues. The removal of corn stover will cause a reduction of topsoil, and nutrients and organic carbon in the soil may also be removed, increasing the need for artificial fertilizer. Switchgrass has relatively low fertilizer and water demand comparing to other fuel sources. It is a perennial that can grow in marginal soil, which also avoids indirect effects of LUC and reduces competition with food crops. Feedstocks and technology of the cellulosic pathway are not yet fully commercialized; the result of LCA variability is still significant due to the improvement in technology over time. Research for cellulosic ethanol largely focuses on improving pretreatment and saccharification process.

The algae biodiesel pathway offers a large potential for biofuel production with high growth rates, photosynthesis efficiency, and lipid content. Algae avoid LUC, and may avoid competition with regular food stock production, but this technology is in its earliest stages.

Sugarcane ethanol has decent amount of production and outperforms other first generation ethanol pathways. The LCA for this pathway is more conclusive than others with more data available, and the variability is mostly driven by indirect factors, LUC, and modeling potential revolution in production. This is the only mature pathway showing significantly less variability with few technology modifications and retrospective analyses possibilities.

Liquid Fuels from Carbon and Hydrogen

Most technology development of alternative fuels production focuses on photosynthetic routes due to the endless supply of light and straightforward extraction of energy from it. However, there is also a wide range of processes and products for non-photosynthetic pathways in renewable fuel production. A class of these, electrofuels, can utilize an extensive range of microorganism and energy sources to produce variety of fuel types. Hawkins *et al.* (2013) discussed the key process and elements of carbon fixation for electrofuel production. The current blueprint of electrofuel projects focuses on autotrophic microorganisms using carbon fixation pathways to consume CO_2 directly for the production of energy-dense liquid fuels. Electrofuel production requires complementary expertise of multiple fields such as synthetic biology, metabolic engineering, and microbiology to produce the desire type of fuel from the CO_2-fixing species. Detailed biochemical characterization of each autotrophic species with cycle type, target product, and enzyme varieties can help to improve the efficiency of fuel production.

Electrofuel production is the process of using non-photosynthetic microorganisms to convert CO_2 directly to energy-dense fuel. There are multiple sources of the electrons needed to power the electrofuel process, including H_2, formate, carbon monoxide, and electricity. In the paper, the authors focus on the use of hydrogen gas as source of reducing power for CO_2 fixation, with hydrogen acting as a reducing agent for electron carriers and production of pyruvate. Though both aerobic and anaerobic process the microorganisms produce storable fuel.

There are currently six different biological pathways occurring spontaneously for carbon fixation. Each pathway has unique features arising from its molecular and biological context, and they differ in efficiency. One of the most common types of CO_2-fixation is the Calvin-Benson-Bassham (CBB) cycle found in plants, algae, and many bacteria. For example, *Ralstonia eutropha* is a metabolically diverse autotrophic bacterium that can grow on CO_2 and H_2; it is able to store excess carbon as polyhydroxyalkanoates (PHA). Wood-Ljundahl (W-L), 3-hudroxypropionate (3HP), 4-hydroxybutyrate (4HB), reductive tricarboxylic acid (rTCA) and dicarboxylates (DC) are also various pathways of CO_2-fixation. One important characteristic to the various CO_2-fixation pathways is the tolerance of the enzyme and redox carrier molecules to oxygen. Some pathways are found in anaerobic organisms, and these pathways utilize oxygen-sensitive reduced ferredoxins as electron carriers.

The efficiency of CO_2-fixation depends on multiple factors. On average, it is found that the W-L pathway and rTCA cycle were the most energy efficient routes, while CBB comes in as least efficient. One of the analyses of carbon fixation reaction is examins the ATP requirements and the reduction potential of electron carriers. For example, ferredoxin has much lower reduction potential than NADPH, but the extra energetic contribution from the difference is insignificant compared to the large difference in ATP cost for pyruvate formation between cycles. Another analysis looked into the individual reactions in carbon fixation pathways. There are many half-reactions in the process of carbon fixation, and some of the processes like carboxylation and carboxyl reduction reaction are energetically unfavorable. To increase the efficiency and reduce ATP requirements, some pathways manage to couple unfavorable reactions to exergonic reactions other than ATP hydrolysis. One of the faster ways to speed up a reaction is the use of enzymes; combining enzymes in novel ways may improve the efficiency of the pathway or reduce the energetic cost. The conversion of CO_2 and hydrogen to electrofuel is also dependent on the use of enzymes, carbonic anhydrase, and hydrogenase.

Hydrogenase catalyzes the reversible conversion of molecular hydrogen and protons in the presence of an electron carrier. Based on the molecular structure, hydrogenase can be categorized: [NiFe]-hydrogenase, [FeFe]-hydrogenase, [Fe]-hydrogenase. Different microorganisms appear to utilize different type; [NiFe]-hydrogenase is widespread among bacteria and archaea, [FeFe]-hydrogenase is found in anaerobic bacteric and eukaryote, and [Fe]-hydrogenase is found only in certain archaea. In CBB cycle hydrogenase helps catalyzing the oxidation of hydrogen for the reduction of $NADP^+$, where in 3Hp and 4HB cycle hydrogenase catalyze the oxidation of reduced ferredoxin and generates a gradient that is used for ATP synthase. Other than the use of hydrogenase, enzymes that enhance the concentration of CO_2 also play a large role in carbon fixation.

Any fuel or organic product from CO_2 that relies on carbon fixation depends on the concentration of CO_2 in the environment. Thus the mechanism of increasing the concentration of CO_2 has developed to compensate the low ambient CO_2 concentration. CBB cycles utilizing RubisCO, 2 CO_2-fixing enzyme that has low attraction for CO_2 and does not differentiate well between CO_2 and the competing substrate O_2, especially need this distinct mechanism to avoid unfavorable loss of carbon in photorespiration. Carbon concentrating mechanisms have only been described for the CBB cycle because 3HP/4HB uses bicarbonate (HCO_3) instead of directly using CO_2. However, exploitation of carbon fixing enzymes for microbial eletrofuels can improve the engineering of enzymes.

The key to carbon fixation is the electrofuel-processing microorganism. The host development requires genetic engineers as well as knowledge of gene regulation and metabolism in the target host cell. Electrofuel is promising relative to inefficient photosynthetic biofuels, but only a handful of electrofuel organism are developed or reported.

Use of Xylose in Biofuel Cell

Because of the depletion of fossil fuels and the increasing awareness of environmental problems caused by them, biofuels have been the focus of much research. A highly specialized type of biofuel uses enzymes and microorganisms in device called a biofuel cell to convert glucose or fructose directly into electricity. Xia *et al.* (2013) discussed the possibility of utilizing xylose as an energy source for biofuel cells and tested the efficiency of the cells. Xylose is an important monosaccharide in the cellulose fermentation industry, comprising almost 30% of lignocellulose materials. In this study, the authors use enzymatic bioelectrocatalytic oxidation of xylose in a biofuel cell rather than breaking it down with a thermochemical method or with microorganism fermentation. To overcome the poor stability of the purified xylose dehydrogenase in ambient conditions, the authors bound it to bacteria and stabilized it by binding the bacteria to the electrodes. The maximum power density and the open-circuit potential of the cell are 63 μWcm^{-2} and 0.58 V.

Biofuel cells have been widely studied utilizing various energy sources for energy production. The authors utilized xylose, an important monosaccharide that has not been commonly used for biofuel cells, in enzymatic biofuel cells. Conventionally, the conversion of xylose to fuel can be achieved by thermochemical methods or microorganism fermentation. The thermochemical process requires harsh conditions, such as high temperature, that are energy consuming; microorganism fermentation processes have very poor yield, and might lead to production of various by-products, which will lower production efficiency. The authors utilized enzymatic bioelectrocatalytic oxidation of xylose in the biofuel cells while facing issues of poor stability and lack of durability. They figured that immobilization of XDH-bacteria on nanomaterial modified electrodes could increase the enzyme activity and improve the stability problem.

There are modifications on the bioanodes and biocathodes of the cells; both biocathodes and bioanodes are pretreated with BOD and other buffer solutions. First the electrodes are prepared with glassy carbon materials surrounded with multiwall-nanotubes; then soaked up in PBS buffer at 0.85 V to cover with poly brilliant cresyl blue (PBCB), then covered with immobilized XDH-bacteria. The measurements of the control group are performed in 0.1 M PBS buffer containing 30 nM xylose and 10 nM NAD^+.

Generally, for a given fuel assembly, the open circuit potential is determined by the difference between the onset potential for catalysis at the bioanode and biocathode; to increase the open circuit potential, the onset potential needs to be decreased. However, the result shows that the onset potential was increased after the immobilization of XDH-bacteria from 0.3 V to 0.4 V. Comparing the result to the NADH oxidation, this increase might be influenced by the morphology of the nanomaterial on the electrodes. PBCB redox polymer covering the electrode was used as the mediator for the oxidation of NADH. The results show that an electrode covered with PBCB is much more negative in oxidation potential than non-covered ones, which indicates an increase in the efficiency of electrocatalytic activity. The concentration of xylose also affects the efficiency of the cell; electrocatalytic current improved dramatically with an increase in xylose concentration from 5 nM to 30 nM. The current reached to 0.59 mA cm^{-2} at 0.2V while the increase reached a plateau after 30 nM. Electrodes were prepared and coated with BOD which, unlike other multi-copper oxidases, works well under neutral pH condition where XDH-bacteria can retain their enzyme activity. Other than the modification of the electrodes, the cells were also tested under various conditions. In the presence of O_2, cathodic current appeared to increase where as there is no increase in current under N_2 saturated conditions. For the cell modified with XDH-bacteric, PBCB and under 30 nM concentration of xylose reached a maximum power output of 63 μW cm^{-2} at 0.44 V.

Microfluidic Biofuel Cell Modifications

In developing various types of biofuel cells, options include the fuel source, the enzymes used, and the interior dynamic flow. Microfluidic cells are structures using the property of laminar flow to prevent the mixing of fuel and oxidant streams. Gonzalez *et al.* (2013) experimented with a microfluidic enzyme-based cell featuring pyrolyzed photoresist film (PPF) as an electrode made on silicon wafers encapsulated into a double-Y-shaped microchannel. There are two common problems challenging enzyme-based systems: short enzyme lifetime and poor electron transfer rate to solid electrodes. To solve this problem, the authors immobilized the enzyme on the electrode surface and substituted the material of the electrodes. The architecture of the cell reduces internal ohmic losses and allows the cell to work in different pH without reagent crossover issues. The cell performance in this study obtains a maximum open circuit voltage of 0.54 V, maximum current density of 290 μA cm^{-2}, and a maximum power density of 64 μW cm^{-2} at room temperature under flow rate of 70 μL min^{-1}.

Microfluidic biofuel cells are a young technology in the biofuel research field; this study examines a microfluidic biofuel cell that is enzyme-based and utilizes glucose and O_2 as reagents. Compared to cells featuring bioelectrodes, microfluidic fuel cells with enzymes in solution have been reported to yield significantly higher power density; however, bioelectrode cells are more cost effective due to the lower amount of enzyme required to run the cells. The design of microfluidic cell in this study features multiple modifications in an attempt to improve power density and relieve common problems seen in microfluidic cells.

As in other biofuel cells, the electrodes were pre-treated with enzyme solutions; immobilizing the enzyme increases the lifetime and decreases the cost of the cells. The electrodes were made of PPF synthesized on a silicon wafer, similar to glassy carbon but less costly and extremely flat; PPF can also be shaped into wide range of structures using photolithography. Except for the silicon chip featuring the electrode, the entire cell is fabricated with laminated polymers compatible with low-cost and mass manufacturing techniques. PPF electrodes were cleaned and soaked in pre-treated enzyme solutions. The PPF bioanode was covered with glucose oxidase solution and the biocathode was prepared with laccase buffer solution. The cell comprises several layers of different laminated materials with a double-Y-shaped channel on one of the layers. The combination of materials provides an optimal fit for the silicon chip containing PPF electrodes.

The first test result is a verification of the suitability of PPF as an electrode material. The result shows that although the current levels of glassy carbon (GC) and PPF electrodes are slightly different, both electrode materials behave similarly in peak potential. The PPF anode shows a higher current density of 460 $\mu A\ cm^{-2}$ compared to 321 $\mu A\ cm^{-2}$, but the PPF cathode passed somewhat less current, 113 $\mu A\ cm^{-2}$, compared to 165 $\mu A\ cm^{-2}$. The authors concluded that PPF is a suitable electrode material in this case. The authors also tested the voltage of PPF electrodes under aerobic conditions compared to an O_2 saturated environment absent glucose. The result suggests that electron transfer is quasi-reversible and that oxygen reduction at the cathode is likely the limiting process of cell performance. The concentration of dissolved oxygen in water is much lower than the concentration of glucose in anolyte; increasing the concentration of oxygen above its solubility will damage the enzymes.

Laminar flow occurs inside the cell so the anolyte and catholyte solution don't cross over, and thus controlling the flow rate of solutions can adjust the cell activity. The result shows that the power density reaches a plateau of 60 $\mu W\ cm^{-2}$ at a flow rate of 40 $\mu L\ min^{-1}$. There are two major performance losses in microfluidic cells: diffusional crossover and depletion zone. In this study, the depletion zone of oxygen at the cathode is significantly limiting the power density at low flow rate. Increasing the transport of oxygen to the cathode is useful until its supply rate equals the turnover rate of the immobilized laccase. Cross-diffusion of fuel and oxidant is less likely to play a significant role due to the shape of the cell. The authors calculated the inter-diffusion zone and concluded that it is reasonable to disregard the possible effect of cross-diffusion. Oxygen depletion at the cathode is likely to be the limiting factor in this system; however, the presence of sufficiently high oxygen concentration at the anode could disable the fuel cell because oxygen is the natural substrate of glucose oxidase.

The best results were obtained with a flow rate of 70 $\mu L\ min^{-1}$ showing a high open circuit voltage of 0.54 V, maximum power density of 64 $\mu W\ cm^{-2}$ and current density of 290 $\mu A\ cm^{-2}$. However, the cell has low stability; the performance dropped over 50% after 24 hours likely caused by the lifetime of enzyme. So this type of cell is more suitable as a disposable unit.

Long-Haul Truck powered by a Hydrogen Fuel Cell

For the past few years, battery electric vehicles and biofuel-based internal combustion engines (ICEs) have been invented and widely modified, and engineers are looking for non-polluting sources of fuel for powering the vehicles. ICEs running on hydrogen have low carbon emissions and do very little harm to the environment. Nonetheless, diesel engines, which are usually used in trucks, running directly on hydrogen can produce more nitrogen oxide (NO_x) than when running on regular petroleum diesel. A solution to the emission of NO_x is powering the truck engines with a hydrogen battery system. A battery fuel system not only reduces the emission of pollutants, it is also potentially twice as efficient and occupies less space and weight. One previous example of hydrogen fuel cell truck can cover 320 km over an 8 hour shift carrying up to 40 kg of a hydrogen on board. While the tank can be refilled in short period of time, the initial cost of the truck is 2.5 times that of a standard diesel truck but saves 30–40% on fuel cost. Shabani *et al.* (2013) designed and analyzed the application of a hydrogen fuel cell to a scale-model truck. The truck was a $1/14^{th}$ scale battery-based replica of a Scania R470 Highline truck with the unique design of a two fuel cell stack and bottled hydrogen fuel. The hydrogen fuel cell system was found to be more efficient than the original batteries; the gravimetric energy density of the fuel cell system was 30% better than the original battery system.

A very large portion of the global energy demand is met through combusting fossil fuels. Transportation especially consumes almost half of the oil production and the amount is consumed rapidly growing. The major issue of fossil fuel based transportation is the depletion of fossil fuel and greenhouse gas pollution, following with the additional challenge of affordability as the quantities of fossil fuel inevitably decreasing.

First to assure the proper operation of the truck, the minimum required power density and current must be reached. For a small-scale replica that originally runs on a 7.2 V, 2400 mAh Li-ion battery with maximum required power of 60 W, the authors substitute two 30-W fuel cell stacks with 7–12 V output that consumes 0.4 bar hydrogen pressure as input. Total weight of the fuel is one of the key elements that affects the efficiency of the model. Metal hydride was selected as the solution for hydrogen storage; each bottle with about 0.5 bar and 1.5 g of hydrogen weigh 165 g and could be used without supplementary heating. The truck carries four bottles of hydrogen.

One of the major issues for fuel cells is the buffer system. An initial performance test showed periodic interruption in the power supply to the motor during hydrogen purging, specifically when the two fuel cells underwent purging at the same time. To improve the stability of the power source, buffer energy was added. An ultra-capacitor was chosen over a Li-ion battery due to its higher energy efficiency. The buffer system is designed so that during the purging period, the charged buffer capacitor is able to support the system absent fuel cells for at least 10 seconds before the next cycle of purge. The fuel cells were able to charge the capacitor in less than 15 seconds.

Compared to the original battery source, the hydrogen fuel cell modified truck improved significantly in gravimetric energy density and duration of run. The scale model was able to run 18 km in about 2.6 hours where as the original battery could only run 1.6 km for about 14 minutes. A record of power output over 5 minutes shows sharp minimum and maximum spikes indicating sudden variation of the fuel cell load relating to the solenoid valve that controls the hydrogen flow. There was a smoother performance with the buffer than without it, but adding the buffer caused a small drop in net power output because of the energy lost within the ultra-capacitor due to its internal resistance in high current.

The experiment shows promising results for trucks running with hydrogen fuel cells. The main weakness of this experiment is the size of the model; which may not scale to a full size truck.

Improving Biofuel Cell Modification

Biofuel cells transforming biological fuels such as ethanol or sugar into electricity are safe and ecofriendly sources of power. However, biofuel cells are often limited to low voltage and are insufficient to provide necessary power for daily use. Like other traditional electrolyte batteries, stacking up the biofuel cells may boost their single cell voltage to an usable level. Miyake *et al.* (2013) performed experiments on multiple ways to improve the voltage of biofuel cells utilizing fructose as the energy source. When stacking cells, each cell has to be isolated with proper wrapping to prevent short-circuits. The authors layered the biofuel cells with enzyme-modified carbon fabric strips and hydrogel sheets to ensure the ion-conduction between the anode and cathode fabric layers; the hydrogel sheet also served as a fuel tank. The modification effectively improved the performance of both bioanodes and biocathodes and maximized cell power to 0.64 mW at 1.21V.

To increase the efficiency and the power of biofuel cells, the authors made three modifications to the cells to prevent short-circuits due to ion-conduction. The first modification was the preparation carbon fabric anodes, which are multi-walled with carbon nanotubes to increase the reaction surface area. They are heated in 400°C and immersed in multiple solutions such as D-fructose dehydrogenase to increase the efficiencies. The result of this modification is significant: it almost doubles the current density. In the first modification, the authors also found that the performance of cells is significantly affected by buffer concentration; buffer is added to stabilize the local pH level change caused by the oxidation process. With a stabilized pH level, enzymes in the carbon fabrics perform with the highest efficiency with a 0.5M buffer with maximum current produced at 0.6V of 15.8 mA.

The second modification is the gas-diffusion of carbon fabric cathodes. This process followed the process used for biliruben oxidase (BOD) cathodes. BOD can catalyze the reaction of O_2 to H_2O without election transfer mediators. The cathode is also treated with heat and multiple solutions including BOD and a surface coat of carbon nanotube solution to make it hydrophobic. To test the effect, the electrode strip was put in an oxygenic pH 5 buffer testing the electric potential versus the current capacity. The performance of a BOD-modified strip reaches to about 1.9 mA cm^{-2}; the additional carbon nanotube coating onto the BOD-modified cathode strip was enhanced to 2.9 mA cm^{-2}. The hydrophobic carbon nanotube coating controls the penetration of excess solution into the carbon fabric electrodes allowing the conduction to optimize. The authors also conclude that control of the buffer concentration may optimize the performance with maximum current of 4.6 mA cm^{-2} at 0V using a 0.25M buffer solution utilizing an oxygen supply from the ambient air through the carbon fiber.

The third modification is through double-network hydrogel films that contain fructose. This modification is prepared through a three-step process: first, the formation of one layer hydrogel film, then another layer of film, lastly with loading of 500 mM fructose. The hydrogel film is later treated with three stock solutions to secure the fructose solution in the film. Both cathodes and anodes went through the process of lamination with double-network hydrogel sheet; the lamination provides the cells with moisture, fuel sources, and buffering for the reaction. The cells are tested with 0.74V, which is about the electric potential difference between fructose oxidation and oxygen reduction. The performance of the biofuel cell is fairly good; it reached a maximum power density of 0.95 mW cm^{-2} at 0.36 V. However, the stability of the cell decreases drastically after a few hours due to drying of the hydrogel. More importantly, the authors found that bending the cell sheet into a cylinder effectively increases the performance of the cell. The laminated bent cell produced a maximum power of 0.64 mW at 1.21V, which is sufficient to light an LED unit. With these types of modification, we may expect a more powerful biofuel cell in the future.

Analysis of Future Biofuel Policy in Thailand

With the high cost of petroleum fuel sources, some developing countries such as Thailand are seeking development of less expensive renewable energy sources, particularly biofuels. Wianwiwat and Adjaye (2013) developed a computational general equilibrium (CGE) model of the Thailand economy that analyzes the government's recent renewable energy 10-year development plan. The policy is a series of bio-liquid fuel promotions that discontinue gasoline-91 use, replace E10 gasohol with E20, replace B3 with B5 biodiesel, and introduce B20 as an alternative diesel. The policy goal was set to replace 25 % of fossil fuel by 2021, in which ethanol use must increase by 262 %. In the short run, the price of fuel shoots up by more than 400 % due to the imbalance of demand and supply; in the long run, the price is stable and the ethanol use almost achieves the goal with a 253 % increase. Surprisingly the development plan does not have a significant affect on the food security.

With the constant rising price of gasoline, the Government of Thailand introduced policies to promote renewable energy and approved a 10-year alternative energy development plan. The plan focuses on increasing domestic renewable energy use to 25,000 kilotons to replace 25 % of fossil fuel use. Although there have been a number of CGE studies for Thailand, none of them have seriously considered energy-sector details and energy policies. The authors mainly want to present a CGE model for Thailand that features various enhancements in the energy sector and use the model to investigate the impact of biofuel-promotion on the economy. The result can be used to predict and develop a plan for achieving economic sustainable goals and improve food and energy security.

The model is based on neo-classical assumptions about economic agent behavior as well as the production and consumption structure. To achieve equilibrium in the model, consumers will maximize utility to their budget constraints, and producers will minimize their cost to consumption constraints. For the convenience of the simulations, the authors divided 51 industries and 62 commodities and electricity sectors into groups; essentially, the more services are aggregated, the less is the degree of substitution between them. The methodology of the CGE model is based on various economic equations and assumptions. The authors also include models of energy composite structure, household demand, and government budget. There are two time periods; the short-run solution reflects a period of about two years or less, while the long run solution is more than five years. There are significant differences in results of the long run and the short-run caused by variables such as the real wages and stock in the palm oil sector. In the short run, land mobilization can only be between three similar crop industries: cassava, sugarcane, and rice; oil palm takes about five years to become a commercial harvest. To achieve the set target of this plan, the use of ethanol must increase by 262 % and biodiesel use must increase by 249 % by 2021.

In a short-run, the aggregate investment and real wages are held constant. The policy has a moderately negative macroeconomic impact; GDP contracts by 0.44 % due to the decrease in government consumption and household consumption. There is also a decline in government revenue from the lost of fuel tax due to the discontinuing of gasoline-91. The positive consequence of discontinuing gasoline supply leads to increase in ethanol output and a doubling in gasohol use. However, the shortage of fuel stock in the short-run causes the price of ethanol to shoot sky-high with a 471 % increase along with a 22 % increase in gasohol price. As a result of enforcing B5 as standard diesel, biodiesel use increases by 117 % leading to a 46 % increase in its price. Lowering the demand for transportation overall the promotion policy did not impact the food sector in the short-run.

There is a slight positive macroeconomic impact in the long-run result. GDP increased by 0.18 % because aggregate investment increases by 2.25 %. In addition, there is a smaller decline in government consumption compared to the short-run resulting from a lower CPI, even though there is a greater decline in total government revenue from tax. The result shows significant increase in biofuel use in the long-run; gasohol increasing by 137 %, ethanol by 253 %, and biodiesel by 126 %. All prices of fuels are cheaper and more stable than they are in the short-run; and apparently the food production is also not affected in the long-run.

The study has shown that the introduction of B20 and enforcement of B5 is unlikely to be adequate to achieve the of 249 % increase; thus it is suggested that the government should enforce higher biodiesel-content fuel such as B7 as standard diesel. The analysis of the model suggests that the plan may advance the economic and energy structure of the country when running for a longer term. Due to the potential short-term adverse impacts on some sectors, it would be more prudent to implement the policy measures in phases.

Conclusions

With a need of alternative energy source, the research on biofuel has made good progress. Biofuels are not only getting cheaper and easier to obtain than fossil fuels, they are also ecofriendly. Although there are still many things that could be improved and modified by future research, biofuel appear to be a viable alternative energy source that might take over from petroleum fuels.

References cited

Borak, B., Ort, DR., Barbaum, J.J., 2013. Energy and carbon accounting to compare bioenergy crops. Current opinion in biotechnology 24, 369–375.

Gonzalez-Guererro, M.J., Esquivel, J.P., Sanchez-Molas, D., Godignon, P., Munoz, F.X., Campo, F.J., Giroud, F., Minteer, S.D., Sabate, N., 2013. Membraneless glucose/O_2 microfluidic enzymatic biofuel cell using pyrolyzed photoresist film electrodes. Lab Chip 13, 2972–2979

Hawkins, A.S., McTernan, P.M., Lian, H., Kelly, R.M., Adams, M.W., 2013. Biological conversion of carbon dioxide and hydrogen into liquid fuels and industrial chemicals. Current opinion in biotechnology 24, 376–384

Kendall, A., Yuan, J., 2013. Comparing life cycle assessments of different biofuel options. Current Opinion in Chemical Biology 17, 439–443

Miyake, T., Haneda, K., Yoshino, S., Nishizawa, M., 2013. Flexible, layered biofuel cells. Biosensors and Bioelectronics 40, 45–49.

Shabani, B., Andrews, J., Subic, A., Paul, B., 2013. Novel concept of long-haul trucks powered by hydrogen fuel cells. Lecture Notes in Electrical Engineering 192, 823–834

Wianwiwat, S., Asafu-Adjaye, J., 2013. Is there a tole for biofuels in promoting energy self sufficiency and security? A CGE analysis of biofuel policy in Thailand. Energy Policy 55, 543–555

Xia, L., Liang, B., Li, L., Tang, X., Palchettic, I., Mascini, M., Liu, A., 2013. Direct energy conversion from xylose using xylose dehydrogenase surface displayed bacteria based enzymatic biofuel cell. Biosensors & bioelectronics 44, 160–163

5. Alternative Energy Efficiency, Storage, and Transportation

Allison Kerley

This chapter reports on a sample of discoveries made in alternative energy research over the course of 2013, with significant findings in the area of solar energy, energy storage, hydrogen generation, and natural gas transportation. Zhou *et al.* (2013) found a way to create an organic solar cell which could be recycled; when immersed in distilled water at room temperature, the CNC film dissolves, allowing for the major components to be separated via a simple filtering process. Bernardi *et al.* (2013) demonstrated the feasibility and efficiency of ultrathin graphene and transition metal dichalcogenides (TMDs) active layers; at 1 nano-meter (nm) thick, graphene and the three TMDs were found to achieve one order of magnitude higher sunlight absorption than the traditionally used GaAs and Si. Li *et al.* (2013) demonstrated the effectiveness of PMDPP3T, a new small band gap semiconducting polymer, in single, tandem and triple junction solar cells. PPMDPP3T was found to absorb up to 960 nm, absorbing almost into the infrared. Yang *et al.* (2013) and Slocum *et al.* (2013) both published works contributing to current energy storage research. Yang *et al.* (2013) discussed their new proof-of-concept lithium/polysulfide semi-liquid battery, a simplified version of a flow battery, as a potential solution to large-scale energy storage. Slocum *et al.* (2013) reported on their proof-of-concept design of a device to store energy produced by offshore wind farms through the use of a pressure pump system. Wang *et al.* (2013) explored the feasibility of a self-powering photoelectrochemical-microbial fuel cell (PEC-MFC) hybrid device to generate hydrogen. Fulke *et al.* (2013) discussed the potential of wastewater stabilization ponds as a potential source of biofuel and CO_2 sequestration. Sa *et al.* (2013) investigated the effects of amino acids as Kinetic Hydrate Inhibiters (KHIs) on the initial formation, the continued growth, and the structure of hydrate blockages in a model CO_2 experiment, with implications for natural gas and oil pipelines.

Allison Kerley

Recyclable Organic Solar Cells on Cellulose Nanocrystal Substrates

Flexible recyclable solar cells are the most recent addition to solar cell research, and have the potential to lower the costs of solar cells while eliminating the need for the petroleum-based components of traditional solar cells. Zhou *et al.* (2013) studied the power conversion efficiency, the rectification in the absence of light, and the recyclability of polymer solar cells created with cellulose nanocrystal (CNC) substrates. CNC's are extracted from plant fibers, and are more environmentally attractive, due to their ability to be recycled, than their petroleum-based counterparts. Zhou *et al.* found the power conversion efficiency (PCE) of solar cells on CNC substrates to be noticeably higher than previous attempts at solar cells on paper-like substrates. The device also showed a low reverse saturation point. The CNC substrates and solar cells were also found to be separable when immersed in distilled water at room temperature, which allowed for the various components to recovered and recycled. A thin Ag layer served as the bottom electrode, whith a MoO_3/Ag layer as the top electrode.

Zhou *et al.* found their CNC substrate-based solar cells to have a 2.7% PCE; they attribute the low PCE to both the limited transmittance of the thin Ag layer, which served as the bottom electrode, and to the uneven and random distribution of the CNC in the clear CNC film. The CNC film is desired to be as transparent as possible while intensifying the light onto the detector. However, the inconsistencies in the spread of the CNC (which are only a few hundred nanometers long) cause scattering of the light, lowering the intensity of the light hitting the detector. The efficiency of the solar cell is determined by the intensity of the light hitting the detector, and the transmittances of CNC substrates were found to be lower than that of glass.

The other piece of the study examined the recyclability potential of the CNC substrate solar cells they created. It was found that the CNC substrate solar cells could be separated into their major components (substrate, organic and inorganic materials) when immersed in distilled water at room temperature then run through several simple filtering processes. When the solar cell was immersed in in water, the CNC substrate disintegrated, allowing the solid wastes to be filtered out with a simple paper filter. The photoactive layer could then be separated from the electrodes by rinsing the solid wastes over filter paper using chlorobenzene, leaving the materials which served as electrodes behind.

Zhou *et al.* theorized that the CNC substrate could be modified to decrease the scattering of the light, increasing the PCE of the solar cell. In addition, they proposed that a different or modified material for the bottom electrode could increase the transmittance.

Extraordinary Sunlight Absorption and One-Nanometer-Thick Photovoltaics using Two-Dimensional Monolayer Materials

Bernardi *et al.* (2013) investigated the absorbance of graphene and three different monolayer transition metal dichalcogenides (TMDs)—MoS_2, $MoSe_2$, and WS_2—alone and in various combinations as the active layer in ultrathin photovoltaic (PV) devices. In calculating the upper limits of the electrical current density (measured in mA/cm^2), each material can contribute to the total absorption of a device. The authors found that subnanometer thick graphene and TMD monolayers can absorb the equivalent short-circuit currents of 2–4.25 mA/cm^2, while 1 nm thick Si, GaAs, and P3HT (commonly used materials in current PV devices) were found to generate currents between 0.1–0.3 mA/cm^2. Further testing suggested that the high absorption of the monolayer MoS_2 is due in part to the dipole transitions between localized *d* orbitals (which were estimated to contribute 2–5% of the total visual absorption) and escitonic coupling of dipole transitions (which were estimated to contribute 5–10% of the total visual absorption). The authors also compared the absorption of TMD monolayers and graphene against their bulk counterparts, bulk TMDs and graphite respectively. Both grapheme and MoS_2 monolayers were found to have absorption values higher than their bulk counterparts by a factor of 2–3. To further examine the applicability of subnanometer thick TMDs and graphene as the active layer in PV devices, the authors examined the performances of a hypothetical device with a bilayer composed of two stacked monolayers of MoS_2/graphene as the active layer. Their calculations found the maximum short-circuit current to be 4.3 mA/cm^2 and the maximum open circuit voltage to be approximately 0.3 eV. They also examined the feasibility and performance of a hypothetical 1 nm thick PV device with an active layer composed of a monolayer of MoS_2 stacked on a monolayer of WS_2, and found the maximum short-circuit current to be approximately 3.5 mA/cm^2 and the maximum open circuit voltage to be approximately 1 V. The authors found that a single TMD monolayer with a subnanometer thickness can absorb as much sunlight as 50 nm of the commonly used Si.

The authors note that while it is not, strictly speaking, possible to "define a macroscopic absorption coefficient in the layer-normal direction for a single layer of MoS_2. By definition this quantity should be averaged over several unit cells of the material", so they calculated the equivalent absorption coefficient α for a monolayer of MoS_2 using the equation $\alpha = A/\Delta z$ where A is the absorbance and Δz is the layer thickness in Å. The monolayer MoS_2 was found to have an equivalent absorbance of $1\text{–}1.5\times10^6$ cm^{-1} and graphene was found to have an absorbance of 0.7×10^6 cm^{-1}, while bulk MoS_2 was found to have an absorbance of approximately $0.1\text{–}0.6\times10^6$ cm^{-1} and graphite was found to have an absorbance of approximately $0.2\text{–}0.4\times10^6$ cm^{-1}.

The authors found that in order for a bilayer of MoS_2/graphene to work, as graphene is a (semi)metal and MoS_2 a semiconductor, a Schottky barrier must be formed. Using Density Functional Theory (DFT), the authors computed the work-function value to be $\o_{MoS2} = 5.2$ eV for a monolayer of MoS_2. In addition, the authors note that the maximum short-circuit current for the MoS_2/graphene layer (4.3 mA/cm^2) differs slightly from the sum of monolayer currents found for graphene and MoS_2 individually (5.9 mA/cm^2) and attribute the difference to the use of different theories used to compute the individual and combined currents. The Bethe-Salpeter equation (BSE) was used to calculate the currents for the individual monolayers and independent-particle theory was used to calculate the current for the stacked MoS_2/graphene layer.

The authors found that a bilayer MoS_2/WS_2 interface can realize PV operation by the use of a type-II heterojunction. In addition they predict that due to the interaction between the two TMD monolayers, there would be significant changes in the bandstructure and absorption spectrum of the MoS_2/WS_2 interface as compared to their individual monolayers.

The authors emphasize that although power density is not a conventional unit for judging the effectiveness of PV devices, they it is crucial to understanding the limits in solar cells with the smallest possible thickness, as well as to estimating the energy achievable from a unit volume or weight of an active layer material.

Efficient Tandem and Triple-Junction Polymer Solar Cells

In response to current interest in organic solar cells, Li *et al.* (2013) investigated the effectiveness of a new small band gap semiconducting polymer, PMDPP3T*, in single, tandem and triple junction solar cells. Multi-tandem solar cells have multiple layers made up of different compounds, with each compound absorbing a different portion of the visible light spectrum. In single-junction cells PMDPP3T was found to have an external quantum efficiency (EQE) of up to 7.0% when mixed with [70]PCBM. In single-junction solar cells, PMDPP3T was found to absorb radiation closer to the infrared light spectrum than current polymers, absorbing up to 960 nm. All multi-junction solar cells utilizing PMDPP3 in conjunction with PCDTBT reached efficiencies between 8.4% and 8.9%. In addition, Li *et al.* also found that tandem and triple-junction solar cells, compared to single-junction solar cells of the same layer thickness, were 50–60% more efficient.

In their tandem-junction solar cells, Li *et al.* use PMDPP3T combined with [60]PCBM for one layer, and PCDTBT combined with [70]PCBM for the second layer. It was found that PCDTBT and PMDPP3T have complementary absorption spectra, absorbing a different part of the visible light spectrum with little overlap. In their triple-junction solar cells, the authors used two layers of PMDPP3T combined with [60]PCBM and one layer of PCDTBT combined with

[70]PCBM. The tandem-junction solar cells had a power conversion efficiency (PCE) of 8.9%, and the triple-junction solar cells had a PCE of 9.6%. In comparison to the single-junction solar cells of the same layer thickness, the tandem and triple-junction solar cells were found to have a 50–60% efficiency increase.

*PMDPP3T is shorthand for poly[[2,5-bis(2-hexyldecyl-2,3,5,6-tetrahydro-3,6-dioxopyrrolo[3,4-*c*]pyrrole-1,4-diyl]-alt-[3',3"-dimethyl-2,2':5',2"-terthiophene]-5,5"-diyl]. [70]PCBM is short for [6.6]-phenyl-C_{71}-butric acid methyl ester.

A Membrane-Free Lithium/Polysulfide Semi-Liquid Battery for Large-Scale Energy Storage

Yang *et al.* (2013) discussed their new proof-of-concept lithium/polysulfide semi-liquid battery as a potential solution to large-scale energy storage. The lithium/polysulfide (Li/PS) battery uses a simplified version flow battery system, with one pump system instead of the traditional two. The Li/PS battery was found to have a higher energy density than traditional redox flow batteries, with the 5 M polysulfide solution catholyte cell reaching an energy density of 149 W h L^{-1} (133 W h kg^{-1}), about five times that of traditional vanadium redox battery. The Li/PS battery cells were also found to have a high coulomb efficiency peak around 99% before stabilizing at around 95%, even after 500 cycles. The authors conclude that the Li/PS cells maintain a steady rate of performance after 2000 cycles, and are more cost-effective than traditional flow systems as the Li/PS cells operate without the expensive ion-selective membrane and can be operated at room temperature.

In the Li/PS battery, passivated metallic lithium foils made up the anode, a liquid lithium polysulfide solution* made up the catholyte, carbon-paper acted as the current collector, and a recently developed $LiNO_3$ electrolyte additive was used as a passivation layer. During operation, the polysulfide solution continually flows through the electrode stacks to generate or store electricity. In down time, the catholyte drains back to the reservoir tank. Due to the simple design, the battery could be scaled either up or down with relative ease.

In the electrochemical tests (which were all done in the 2032 coin cell configuration), the initial discharge capacity was found to be 172 mA h g^{-1}, and the second discharge was found to reach a capacity of 295 mA h g^{-1}, corresponding to a capacity of 48 mA h cm^{-3} for the whole catholyte. The average discharge voltage was 2.45 V, and the average voltage was 2.30 V, which indicated that the reaction is highly reversible. In the constant voltage cycling tests, which was investigated at a 0.8 C rate, it was found that the cells without the $LiNO_3$ electrolyte additive could not be properly charged due to a strong shuttling effect, and reached a coulomb efficiency of 15% (as compared to the coulomb efficiency of 99–95% for the cells with the additive). Cells with the additive, however, reached energy densities of 108 W h L^{-1} (97 W h kg^{-1}) for 5 M catholytes and 61 W h L^{-1} (59 W h kg^{-1}) for 2.5 M catholytes. The capacity fading rate was found to be as low as 8.4% (for cells with

the 5 M catholyte) and 5.0% (for cells with the 2.5 M catholyte) per 100 cycles. In addition to tests run on the Li/PS battery, the authors ran tests on effects of various parameters on the cycle life and the coulomb efficiency, which are not covered in this summary.
*Lithium polysulfide (Li_2S_8) in 1,3-dioxolane (DOL)/1,2-dimethoxyethane (DME)

Ocean Renewable Energy Storage (ORES) System: Analysis of an Undersea Energy Storage Concept

Slocum *et al.* (2013) propose a new design for an energy storage and generation unit composed of underwater concrete spheres and offshore wind turbines. The proposed design utilizes pumped storage hydraulics (PSH). During times of low energy demand from the grid, the cylinder would contain water at equal pressure with the surrounding ocean. In the proposed design, the floating wind turbines generate energy and the excess energy is used to pump water out of the storage sphere, creating a vaccum. When energy is needed from the sphere, the turbine would open, allowing water to pass through into the sphere. The proposed sphere design would have an inside diameter of 25 m, and would retain a 1/20th-atm environment when fully discharged. The proposed design could be used without alteration in depths between 200 and 700 m, and would continue to be economically feasible to a depth of approximately 1500 m. The authors tested a small-scale dry version of the proposed design, with the test sphere having an inner chamber diameter of 75 cm, with a ten meter height difference from the top of the pump and wind turbine to the top of the sphere. The test unit was found to have a low round-trip efficiency of 11%, which the authors attribute to their inability to use the most efficient pump and turbine technology due to the small size of their test model. They calculated that in a full scale model, the lowest round-trip efficiency would be 70%.

The steel-fiber reinforced concrete and glass-fiber concrete are indicated as the ideal materials for the creation of the sphere, as both have been shown to minimize cracking under high pressures. The authors also mention that further research is needed regarding corrosion of the pumps, clogging from sediment ingestion during turbine operation, and the effects of and on nearby marine life. However, they also propose the spheres as potential artificial coral reefs and marine habitat. In addition, the authors calculated a conservative 20 year lifespan for the spheres and turbines. However, as various concrete offshore platforms in the North Sea have continued to operate for over forty years, the authors estimate that depending on the materials used, their spheres and wind turbines could last a great deal longer than 40 years. At the end of the lifespan of the pump system, the pump/turbine units and transmission lines would be salvaged, but as the spheres themselves could be left in place as the structural materials are believed to be harmless to marine life. Lastly, the authors found multiple potential sites world-wide with wind speeds necessary to power the turbine, underwater terrain structurally sound enough to hold the units in place, and close enough to population centers for the energy to be able to reach consumers.

Self-Biased Solar-Microbial Device for Sustainable Hydrogen Generation

Most hydrogen generating devices require an external addition of a 0.2 to 1.0 V electric potential in order to sustain the hydrogen generation. Wang *et al.* (2013) explored the feasibility of a self-powering photoelectrochemical-microbial fuel cell (PEC-MFC) hybrid device to generate hydrogen. The PEC-MFC was a combination of a photoelectrochemical fuel cell and a microbial fuel cell. The Hydrogen production of the device was tested when powered by a ferricyanide solution inoculated with a pure strain of *Shewanellla oneidensis* MR-1 and when powered by microorganisms found naturally occurring in the municipal wastewater. In both scenarios, given replenishments of fuel, the device produced enough voltage to be self-sustaining. However, when the device was powered by wastewater it produced both a lower current and a smaller hydrogen production than when powered by ferricyanide solution.

This study primarily examined whether the PEC-MFC device could produce enough power to be self-sufficient and self-sustaining. The PEC device used rutile TiO_2 as the photoanode material, and when run independent of the MFC at zero external bias (no outside input of energy) yielded a photocurrent density of 0.0013 mA cm^{-2}. No gas bubbles were observed in the PEC device when run individually, hypothesized by the authors to be the result of the low current. When the MFC fueled by the ferricyanide solution was run independent of the PEC, generated a current of 0.1 to 0.6 mA, with a spike consistently returning the current generation to peak efficiency immediately following the injection of new fuel and slowly decreasing as the nutrients became depleted. It is unclear whether or not the MFC unit was run without an external source of energy, and no mention of gas accumulation was made by the authors.

When run in conjunction, the PEC-MFC generated a reproducible photocurrent current density of approximately 1.25 mA cm^{-2} at zero external bias, with gas bubbles containing H_2 continuously forming in the device. However, the authors state they were unable to accurately measure the H_2/O_2 ratio due to the limitation of the instruments available. The authors proposed the MFC device as serving as a battery for the hydrogen producing PEC device. In addition, they confirmed that the overall current generation is determined by the MFC performance, which varies depending on level of activity of the microorganisms/bacteria (which were found to be dependent on the amount of nutrients in the surrounding solution).

In addition to using a ferricyanide solution as the catholyte for the MFC device, they also tested the MFC device using air as the cathode with *S. oneidensis* as the anolyte, as well as air as the cathode and microorganisms found in municipal wastewater collected from the Livermore Water Reclamation Plant (Livermore, California, USA) as the anolyte. A 1:1 ratio of anaerobic and aerobic sludge collected from the Livermore Water Reclamation Plant was used to inoculate the MFC until the electrochemically active bacteria were enriched and generated an electrical current. The air/wastewater PEC-MFC device had a current density between 0.0 and 1.4 mA cm^{-2}, whereas the air/ferricyanide PEC-MFC device shows the current density to reside between 0.2 and 1.6 mA cm^{-2}.

Potential of Wastewater-Grown Algae for Biodiesel Production and CO_2 Sequestration

In response to a growing fear surrounding increasing levels of CO_2 in the atmosphere and rapidly dwindling supplies of traditional oil as a source of energy, Fulke *et al.* (2013) investigated the CO_2 sequestration rate (as a source of CO_2 mitigation), the biomass creation (as a source of biofuel), and lipid composition of algae used in the wastewater stabilization ponds of industrial wastewater treatment plants. They found that the green algae species naturally occurring in the wastewater stabilization ponds had a lipid structure equivalent to vegetable oil currently used to produce biodiesel. In the two most dominant algal classes Chlorophyceae and Cyanophyceae, they found four distinct species (*Scenedesmus dimorphus, Scenedesmus incrassatulus, Chroococcus* sp. and *Chlorella* sp.) currently being globally explored as sources of biodiesel. The authors isolated and cultured samples of these four species and examined the biomass concentration, lipid content, and CO_2 fixation rates, finding that the samples where all four of these species were present (as opposed to each species cultured individually) had a biomass concentration (g L^{-1}) and lipid content (g g^{-1}) nearly twice as high as their individual counterparts, and a CO_2 fixation rate (g L^{-1} d^{-1}) at least double individual species cultivations. The authors concluded that industrial wastewater could support a diverse culture of algal species capable of being used as a source of biodiesel.

Fulke *et al.* collected samples of water from ten different locations in the wastewater stabilization pond at a currently active vehicle manufacturing plant in the western Maharashtra region in India. They found 27 species of Chlorophyceae, 16 species of Cyanophycea, 14 species of Bacillariophyceae, 4 species of Euglenophyceae, and 4 species of Chrysophyceae in the wastewater, with a Shannon-Wiener Diversity Index range from 2.91 to 3.66. The Nile Red staining method was used to determine the lipid content and to identify the intracellular lipid content (used in the creation of biodiesel), with cells with higher lipid concentrations being more favorable to biofuel creation. Four of the algae species found (*Scenedesmus dimorphus, Scenedesmus incrassatulus, Chroococcus* sp. and *Chlorella* sp). are currently being globally explored as potential sources of biodiesel, so Fulke *et al.* chose these algae to further investigate the lipid content and biomass creation during stress and no-stress scenarios. They cultivated each of the species individually in the lab over 14 days, each in a culture with abundant nutrients and in a culture with limited nutrients. They found that upon nutrient depletion, the algae had a higher lipid concentration.

Hydrophobic Amino Acids as a New Class of Kinetic Inhibitors for Gas Hydrate Formation

Sa *et al.* (2013) investigated the effects of amino acids as kinetic hydrate inhibiters (KHIs) on the initial formation, the continued growth, and the structure of hydrate blockages in a model CO_2 experiment, with implications for natural gas and oil pipelines. They found that hydrophobic amino acids were more effective KHIs than the more commonly used polyvinyl pyrrolidone (PVP), and that in general, amino acids with shorter alkyl chains were more effective KHIs than those with longer alkyl chains, with glycine and L-alanine being the most effective KHIs. Both PVP and glycine as KHIs caused a decrease in the rate of formation of hydrogen blockage. When comparing amino acids with varying alkyl chain lengths in both fresh and memory water, in both cases as the length of the alkyl chain increased, its ability to act as an effective KHI decreased. The crystal structure of the hydrates formed did not change in the presence of the amino acid KHIs, however, all the amino acids tested, regardless of their hydrophobicity, were effective in inhibiting hydrate blockages once the blockages had begun to form, as seen by the increased number of ice crystals in the hydrate in the presence of glycine.

Using nucleation kinetics measurements to observe the onset of hydrate blockage formation, Sa *et al.* examined the effects of different amino acids and PVP on hydrate formation in fresh water and memory water. The "memory effect" of memory water is a phenomenon in which hydrates form more easily in gas and water that has formed hydrates in the past. In the presence of freshwater, an increase in the glycine concentration from 0.01 to 1.0% mol was found to have no effect on the average subcooling temperature. In the presence of memory water, while PVP did not display any effect on the inhibition of hydrates, glycine (at an increased concentration of 1.0% mol) slowed the formation of hydrates. The authors examined the effects of KHI's on the growth of hydrates after initial formation. At 0.1% mol, glycine and L-alanine considerably lowered the growth rate, while amino acids with longer tails had little or no effect on hydrate growth, and in some cases accelerated it in the early stages of formation. The authors propose that above a critical chain length, hydrate growth inhibition was adversely effected.

Synchrotron powder X-ray diffraction (PXRD) was used to identify the structure of the hydrate blockages, enabling Sa et al. to determine whether KHIs affected the structural makeup of hydrate blockages. It was found that in the presence of glycine, hydrate blockages displayed more ice crystals, which was attributed to water molecules freezing instead of forming hydrates.

Conclusions

While the research isn't there yet, innovations in solar cells discussed in this chapter indicate that flexible, organic, and highly efficient thin solar cells are in the cards. Research in natural gas suggests future innovations increasing the efficiency of transport and generation of fuels alternative to oil and coal. Lastly, innovations in energy storage suggests that wind and wave power will become more competitive in the future, with the ability to store energy and thus feed a more steady stream of energy into the grid.

References Cited

Bernardi, M., Palummo, M., Grossman, J., 2013. Extroardinary sunlight absorption and one nanometer thick photovoltaics using two-dimensional monolayer materials. Nano Letters 13, 3664–3670.

Fulke, A., Chambhare, K., Sangolkar, L., Giripunje, M., Krishnamurthi, K., Juwarkar, A., Chakrabarti, T., 2013. Potential of wastewater grown algae for biodiesel production and CO2 sequestration. African Journal of Biotechnology 12, 2939–2948.

Li, W., Furlan, A., Hendriks, K., Wienk, M., Janssen, R., 2013. Efficient tandem and triple-junction polymer solar cells. ACS 135, 5529–5532.

Sa, J., Kwak, G., Lee, B., Park, D., Han, K., Lee, K., 2013. Hydrophobic amino acids as a new class of kinetic inhibitors for gas hydrate formation. Scientific Reports 3, 2428.

Slocum, A., Fennell, G., Dündar, G., Hodder, B., Meredith, J., Sager, M., 2013. Ocean renewable energy storage (ORES) system: analysis of an undersea energy storage concept. Proceedings of the IEEE 101, 906 – 924.

Wang, H., Qian, F., Wang, G., Jiao, Y., He, Z., Li, Y., 2013.Self-biased solar-microbial device for sustainable hydrogen generation. ACSNANO 7, 8728–8735.

Yang, Y., Zheng, G., Cui, Y., 2013. A membrane-free lithium/polysulfide semi-liquid battery for large-scale energy storage. Energy and Environmental Science 6, 1552–1558.

Zhou, Yinhua, Fuentes-Hernandez, Canek, Kahn, Talha M., Liu, Jen-Chieh, Hsu, James, Shim, Jae Won, Dindar, Amir, Youngblood, Jeffrey P., Moon, Robert J., Kippelen, Bernard. 2013. Recyclable organic solar cells on cellulose nanocrystal substrates. Scientific Reports 3, 1536.

Section II: Marine Resources

6. Marine Fisheries

Neha Vaingankar

Our oceans are a series of large, complicated and integrated ecosystems and we rely on these ecosystems as a valuable resource for food and resources. So it is important for us to keep a close eye on how we are managing our fish stocks and fisheries. With a growing population, this is more important than ever because there are more and more people we need to feed. Especially vulnerable marine ecosystems like coral reefs are critical for fish stocks and scientists need to pay close attention to them.

About ninety percent of fishery catches come from marine fisheries. Most fisheries are located near the coasts since both fish and people are much more abundant near the coastal shelves. Despite their value, marine fisheries have been affected by a variety of adverse factors including overfishing, acidification of the ocean, and climate change. For a variety of reasons, it is in the best interest of a nation to protect its local fisheries. Firstly, they can be important socially as a primary food source. In addition, nations that do a good job of preserving the biodiversity in their oceans attract tourists and ecotourism can be a booming business. Finally, we must all understand that it is our moral duty to protect our oceans. They provide us with so much, and we should do our best to make sure that they remain in good condition.

A variety of management practices has been suggested, ranging from database creation to marine protected areas (MPAs). MPAs are regions where human activity has been restricted in order to protect an ecosystem, its surrounding area and the organisms living there. These can be controlled by local, state, territorial, native, regional or national authorities and can differ nation to nation.

The issue with creating management plans in the ocean is that there is a variety of abiotic and biotic factors that affect how the ecosystem works. It is important to understand the different trophic interactions between species and figure out how changes in population affect the food web. The resources and conditions of an environment also play a large role in the balance of an ecosystem. For example, temperature has proven to be a large player in the way fisheries move through the oceans. The concentrations of the fish stocks change as climate changes so management strategies have to take this into account.

MPAs are only effective if people obey them and if there are means of regulation. However, when those things are in order, the effects are very evident. By closing off areas from fishing, once depleted populations can grow and flourish and global extinctions of some species of fish have been prevented. By prohibiting fishing, or even just banning certain types of fishing gear in certain regions, MPAs have been able to restore biodiversity in some areas. This has proven to be very important especially in places that rely on the ocean for social economic security.

Another tactic that has been suggested for the management of marine fisheries has been to create a data base that compiles a variety of traits of an ecosystem to provide the best management strategy for an ecosystem, keeping track of the movement of populations and changes in the environment throughout different seasons. This task has proven to be easier said than done, as research shows that there are massive amounts of data to be sorted through and collected in order to create the best database.

This chapter aims to look at different MPAs and fisheries databases and see how marine fisheries are affected by climate change. The main purpose is to explore different options for conservation of fisheries and see how and if global warming has an effect on movement of marine fisheries.

Criteria for effective Marine Protected Areas (MPAs)

MPAs are good for management systems but it has been difficult to research the wider scale impacts on fish stocks, ecosystems, and fisheries. MPAs must have a specific size of a closed area and have a level of protection of essential habitats for exploited resources. Integration with other MPAs is also ideal because it creates wider integrated fisheries management plans, and finally, MPAs must have efficient monitoring and regulation systems, including participative decision making that ensures restrictive measures are respected by all commercial and recreational fishers. Originally, systems of management employed a single species approach but recently, management takes into account other species, even non-commercial species, habitats, other sources of disturbance and different anthropogenic pressures. There are no-take MPAs, where absolutely no fishing is allowed, and zoning or time limitation MPAs, where there is limited access to the area. The main purpose for these zones is to preserve biodiversity, conserve sensitive species, protect habitat and contribute to limitations of fishing pressure. Mesnildrey *et al.* (2013) attempt to figure out the impact of MPAs on fisheries and other ecosystems and what the criteria of 'ecological efficiency' are for these management systems.

Mesnildrey *et al.* at the University Europeene de Bretagne compiled different case studies and scientific papers to create a research paper that summed up the most effective management strategies for creating MPAs. Sixteen well-documented MPAs were used in order to cover a wide variety of characteristics from depth, to size, to seafloor and type of organisms pursued. These MPAs were located all over the world. At each of these MPAs, the authors looked at the surface area of the zone, the access regulation (no-take, zoned, partial, etc.), ecological effects inside and outside the MPA, and impact on the fisheries. The authors wrote that it was much easier and more reliable to monitor the no-take zones compared ones with variable restrictions.

First, the authors looked at the ecological effects of fishing resources and the consequences for fisheries. Their research led them to find a positive impact on the fish and ecosystems under fishing reserves, since banning fishing activities allowed protected communities to revive, especially when fishing pressures were high. Another positive effect was that outside the fishing reserves there was a spill over effect where excess fish from the MPA venture outside of the MPA.

Spawning biomass increases under fishing reserves since the larvae are free to develop. However, a negative consequence of these protected areas is that at first, having a fishing reserve leads to lower catches and redistribution of the fishing efforts to other zones is costly for the fishers and can cause congestion of zones remaining open to fishing.

Fishing zones are good because they decrease mortality, especially when fishing reserves cover areas where high densities of exploited species are observed. Because of this, fishers cannot relocate far outside the closed area so they fish right up to its boundaries, a phenomenon called fishing the line. Fish populations in MPAs are more resilient to combined effects of environmental fluctuation and fishing pressure because of their faster reconstruction rates and lower risk of collapse.

The overall benefits of MPAs include the protection of habitats with specific interests for exploited species resources, protection of highly productive areas able to provide significant spillover effects or larval dispersion, and the reduction in overall fishing mortality. Overall, these MPAs aim to protect the variable resources under environmental and anthropogenic pressures and are usually strategically located at geographical hot spots and specific habitats, but the effects of fishing reserves depend on connectivity between closed areas and neighboring zones.

Often times, soon after a marine fishery has closed, the fishery management effects inside the borders can only be noticed after many years of protection so ecosystem restoration must be monitored over several years. The size of the fishing reserve has an impact; the larger the area, the more diversity it can foster and the higher the number of exploited species there will be. When it comes to outside the fishing reserve, the bigger the better, because it is much easier to influence other neighboring zones. Although initially MPAs were mainly used for conserving biodiversity, they have been extremely useful in managing marine fisheries. Mesnildrey *et al.* were able to pick out the best types of MPAs, based on time and size of the area.

Effectiveness of MPAs

Marine protected areas are a major cause of dispute especially in coastal and island regions like Seychelles, off the western coast of Africa. In recent times, tropical regions all over the world have experienced a huge boom in fishing of holothurians (sea cucumbers). Almost all of the holothurian fisheries are considered fully exploited, in decline, or entirely collapsed. The reason for the high demand is for the holothurian's medicinal purposes as well as its supposed aphrodisiac qualities. In many tropical coral reef regions, locals rely on these invertebrates for their livelihoods. However, due to the density-dependent reproduction patterns and late maturing of these organisms, holothurians are very vulnerable to over-exploitation. Many MPAs were established in Seychelles 20 years ago that pre-date the wave of heavy exploitation in current times. Cariglia *et al.* (2013) aims to understand the effectiveness of these MPAs and measure the economic value of these holothurians.

Cariglia and her team at the School of Marine Science and Technology in Newcastle, UK looked at seven regions around the Seychelles islands: Cousin, Mahe E, Mahe NW, Mahe W, Ste. Anne, Praslin NE, and Praslin SW. These regions were chosen because they had been previously used by the Seychelles Fishing Authority for other studies. Within each region, 3 sites were chosen unsystematically, for a total of 21 sites. At each site, 16 count areas were delineated with a seven-meter rope, which acted as a radius for a circular area. The number of holothurians was counted in each circle and each holothurian was identified to the species level if possible. There were a total of 336 count circles total. The holothurians were then grouped into classes, which reflected their current economic value: low, medium and high. Only nine out of the 21 sites were located in MPAs. Cariglia *et al.* also realized that the type of habitat could affect distribution of marine organisms. Benthic composition was estimated and expressed as a percentage of cover.

Thirteen different species and a total of 978 holothurians were observed throughout this one month long study. Certain counts had over 100 holothurians while others had none. However, it is to be noted that counts where more than 15 holothurians were observed were all located in sites protected from fishing. Especially prominent were holothurians of high commercial value in the Cousin region (where fishing is not allowed). Low valued holothurians were found within the Praslin NE, Mahe NW, and Ste. Anne regions, where fishing is prohibited only at some sites. Overall, the probability of observing holothurians in counts in MPAs was twice as high as seeing them in counts subject to fishing. After separating them into their respective economic categories, holothurians with high or medium value were 10 times more likely to be observed in MPAs than areas subject to fishing.

Cariglia *et al.* found that habitat was a determinant of holothurian presence. Occurrence of high and medium value holothurians was associated with rocks and coral habitats. Among low value animals, sand was important in protected areas while rock was important in unprotected area. These results were obtained by using principal component analysis to correlate habitats with MPAs and unprotected areas.

Overall, Cariglia and her team found that the presence of MPAs was a significant determinant of high, medium and low value holothurian occurrence on the reefs. This same trend was noted in other projects, specifically in the Chagos Atolls and Maldives and Reunion Islands. In the discussion, the authors explain that closing fisheries as they expand and develop often leads to little success. The management process is done a little too late. In order to make MPAs successful, information regarding biological, ecological and behavioral patterns are required. Another aspect to consider is the gene flow of the species between MPAs. By having MPAs designed so that holothurians of different regions can interbreed, genetic diversity will be increased and the animals will become more resilient to changes. The main point is that successful MPAs can maintain a population at high densities despite high fishing pressure. All of the benefits of the target organism must be understood, authorities and fishermen must recognize this area, and the economic and ecological advantages should be acknowledged.

Fisher's response to MPAs

Few studies have been done that assess the dynamics of the fishers' adaptations to the loss of fishing grounds, from various management mechanisms. The loss of fishing grounds may affect the distribution of some species, so it is important to understand and include the fishers' behavior in response to these changes. Horta e Costa *et al.* (2013) attempted to be the first to effectively study fishing effort allocation and dynamics of artisanal fishers' by following individual fishers' choices of fishing location and gear type at different stages of the MPA implementation. Gear type, habitat features, and MPA design all influence the fishers' behavior. Interestingly, fishers tend to stay near their original fishing spots, and most other vessels respect that. When fishing efforts are higher, catchability of each gear type may be reduced. This affects the expected benefit from protection. However, fishers with multiple gear types have a better chance of adapting to management rules. Jigs and traps faced smaller reduction of fishing grounds than nets. Tracking of the vessels and gear and finding out what factors affect selection of a fishing ground, may allow for understanding of fishers' choices and adaptations to different situations and the dynamics of these small-scale fisheries.

Horta e Costa and her colleagues at the Instituto Universitario Lisboa, University of Algarve, Universidade de Lisboa, and University California Santa Barbara focused their research on Arrábida Marine Park just west of Portugal. It covers fifty-three square kilometers with rocky shores and adjacent mixed substrates. The management plan includes one fully protected area (FPA), four partially protected areas (PPA), where artisanal fishing with traps and jigs are allowed, and three buffer areas (BA) where licensed fishing vessels and recreational fishing can occur. Commercial diving, spearfishing, trawling, dredging and purse seining are not allowed in any part of Arrábida. These rules were gradually implemented within the first four years of the experiment, starting with openings of the BAs. The FPA was integrated in two parts, the eastern half and then the western, both starting out as PPAs. The FPAs were then fully enforced after all of the real PPAs were created. The sampling was done in five different periods; the first was before implementation, three periods during implementation (year 1-3) and finally one period after implementation. For data collection, Horta e Costa *et al.* surveyed vessels from the top of a cliff early in the morning. They located vessels using a topographical triangulation method, which is very accurate. The three fishing gears they looked for were traps and trammel/gill nets, located by buoy geographical coordinates and jugs, identified by vessel locations.

In looking at the number of fishing traps in the Arrábida Marine Park, they used the location of buoys as an indicator of trap locations. Horta e Costa *et al.* compared the density of the buoys to different things, such as the distance to the main port, depth of the water and distance to the FPA. In the before and year one periods, as the distance from the port increased, the density also increased. In years one through three and the after period, water depth was measured. The density of buoys decreased steeply from about eighteen to twenty meters and increased from eighty to ninety meters. Initially, the density decreased as the distance to the FPA increased, but at around 8000 meters form the FPA border, the density of the buoys increased. Researchers found a higher density of traps in the sand compared to muddier and rocker areas. Traps were found mostly close to homeport. However, depth was the stronger influence on density of buoys. The specific type of habitat the vessels were found in was a sandy and rocky shallow reef, indicating that they were harvesting octopus. In looking at the spatial distribution of the vessels, initially, there were clusters closest to home and port. However, the cluster shifted into another group nearer to the coral reefs.

Nets were also located by identifying the location of the buoys. In the before, year one through two, and the after periods the density of nets increased with distance to the port. The density of nets decreased with the depth of the water up to 20 meters and then increased after that. As the distance to the PPA increases, the density of the fishing nets is also increasing, with a slightly slow start. The same trend was found in FPA, but had no initial decrease in density. The habitats that these vessels were found in were primarily rockier areas, versus muddy and reefs, and sand had intermediate values, indicating target species of cuttlefish, sparids and soles. Buoy density was highest near rocks indicating that the fishers like shallower habitats. There were two main clusters of vessels, one in front and one just west of the port. The third cluster occurred in the initial parts of the study, but most likely merged with other clusters of vessels.

Lastly, jigs were located using the location of different vessels because the jigs are placed directly below the boats. As the distance to the port increased, the density of vessels also increased. As the depth of water increased, so did the vessel density, with an initial drop with in the first 200 meters. Finally, the correlation between the distance to the PPA and vessel density was not very regular. The habitat has more rocky reefs and less sand and mud. Jigging takes place close to the shore, near rocky reefs, targeting mallcephalopods. Jigging is mainly influenced by depth and habitat type. At the beginning, there were three large clusters of vessels but in the end there was a single large cluster in front of a place where no fishing regulations had been enacted. This either meant that the management plan had worked, or some fisheries actually profited by staying away from the shore line, making it possible to capture stray species over the sandy bottom adjacent to the shallow rocky reefs rather than waiting until the species migrated to BAs. Fishers tend to stay as close as possible to their original fishing grounds or nearest to the homeport. That is why it is so interesting to see that there is a change in fishers' behavior after the implementation of the MPA.

Effect of Climate Change on Global Fisheries Catch

Climate change is a cause for concern to many different ecosystems, including marine fisheries. Cheung *et al.* (2013) delves into the effects of rising sea surface temperature (SST) on the mean temperature of the catch (MTC). MTC is an index that represents the average inferred temperature preference of a species weighed by the annual catch. Overall, the results showed that the composition of marine fisheries catch is significantly related to changes in SST. More warmer water species are being caught at higher latitudes and fewer subtropical species are being caught in the tropics. This is cause for concern as many developing costal nations rely on the maritime industry, not only environmentally, but also economically and socially as well.

W.W. Cheung and his colleagues at the University of British Columbia found that the global MTC increased each decade from 1970 until 2006, specifically in non-tropical the northeast Pacific Ocean and northeast Atlantic Ocean. Over these years, global temperature preference increased at a rate of about 0.2°C every decade, and the effects were even more pronounced in non-tropical areas. To quantify this, Cheung *et al.* created large marine ecosystems (LME's) to account for most of the world's fisheries. Overall, 52 were used. Spacing of fisheries depends on the best possible environment for those species to live in, including biotic and abiotic factors. Distribution of some marine fisheries has changed over time due to changes in these ocean conditions. In measuring the MTC, Cheung *et al.* inferred the temperature preference of each species on the basis of its modeled distribution from the "Sea Around Us Project" and the Food and Agriculture Organization's fishery database.

Cheung *et al.* assessed the preference of fish species in different areas by calculating the MTC. Global warming leads to catching more warm water species overall and therefore a higher MTC. Cheung *et al.* found that continents towards the north of the globe, like Europe, North America, and northern Asia all show a higher rate of change in SST with MTC rising at higher latitudes, meaning warm water species are being caught further north. Cheung *et al.* also found that MTC changes and SST changes correlate in almost every LME implying that catching warmer water species is a result of changes in ocean temperatures. Tropical marine ecosystems also correlated to the MTC, which stabilized at twenty-six degrees Celsius.

Because MTC is the main index used in figuring out the effects of climate change on marine fisheries, Cheung *et al.* explains why MTC works. Firstly, in the North Sea, scientists compared the change in MTC from the catch data to that of data collected from deep-sea trawlers that scrape the bottom of the ocean for fish and other marine organisms. This shows that the MTC is consistent regardless of what depths the species are coming from. Another example of why MTC is a valid proxy to examine changes in composition of catches in a region in relation to the temperature preference of the exploited animals is because an increase in fishing conservation efforts is related to the temperature preference of the exploited species. However, the positive correlation between the two was only found in about 19 LMEs, which shows that the MTC trend is not a result of the depletion of large fish by fishing. Next, in the case of misreporting of catch data, initial MTC values were not significant. However, after fixing the catch values, the MTC rates of change became significant. This sort of change in significance was not seen when different SSTs were used to calculate species' temperature preference.

Overall, the results show that the change in composition of marine fisheries catch is significantly correlated with the temperature increases in the ocean. These changes in temperature will have a large effect on the socio-economic situation of poorer countries in the tropics, which rely on these fisheries for their income.

Climate fluctuations on Fisheries in a Sub-Arctic Environment

Climate change influences marine ecosystems in different ways. For example, fishery management plans fail because of unanticipated changes. Intense exploitation of fisheries may lead to bottom-up control of the food chain and greater sensitivity to climate change. Because climate change occurs so slowly, it is difficult for scientists to see the ecosystem impacts right away, but gradually, the effects become evident in the interactions between fishing and environmental variability. In this paper Durant *et al.* (2013) aim to explore the effects of fishing and climate change on the structure of populations of sub-Arctic ecosystems, especially when it comes to temperature fluctuations and fishing-induced changes in spatial and demographic population structure. They are particularly interested in shifts in spatial and demographic population structure that affect the recruitment and population growth rate. The results show some patterns as well as differences in the relative importance of fishing and climate on the populations and ecosystems examined.

Durant and his colleagues at the University of Oslo in Norway compiled five scientific papers in an attempt to understand how climate impacts fisheries in the arctic. To explore the topic of ecosystem impacts of temperature change, their department, the Centre for Ecology and Evolutionary Synthesis (CEES), hosted two international workshops. The workshops brought in scientists from Canada, France, Norway, Russia and the United States to discuss topics of theoretical ecology, animal behavior, fish ecology, fisheries oceanography, and ecosystems modeling. Fish stocks were all from the Arctic Ocean and North Sea. Different tests were done, with varying conditions. The authors say that it is not possible to assume that fish populations display a linear response to climate change and overfishing. Spawning stock biomass, recruitment, selective fishing, and temperature were also taken into consideration when analyzing fish stocks since these variables can change over time.

The first paper looks at temporal shifts and temperature effects on recruitment dynamics. Although density-dependent factors have strong effects on juvenile fish populations, density independent factors were assumed to be of primary importance in regulating pre-juvenile survival. The authors related variation in sea temperature, spawning stock biomass, and mean age in the spawning stock to temporal patterns in recruitment dynamics of 38 commercially harvested North Atlantic fish stocks. They learned that is was important to be aware of over-interpretations of linear environmental-recruitment associations because underlying relations may be non-linear and variable over time. The data suggested that as temperature increased, there was an increase in pre-recruit mortality over time in at least 14 stocks.

The second paper looks at population spatial structure. All animal populations use their physical environment as a gauge for special structure. Changes in this spatial structure can affect the resilience of a population to human and environmental impacts. Changes in species abundance usually leads to large-scale changes in spatial distribution patterns. This can eventually lead to a set of new environmental constraints and interspecific interactions with consequences on a community level. The authors explain that the importance of understanding spatial ecology of populations of fish help us to study mechanisms that can lead to rapid changes in population abundance, tell us the health of the fisheries, and characterize the effects of changing these spatial patterns on the populations susceptibility to human exploitation and environmental variability. They found that populations can be spatially structured, regardless of genetic or demographic connections and that changing these special structures has a negative effect at the population and community levels.

The third paper looks at methods of predicting fish recruitment. The authors looked to see if spawning stock biomass (SSB) was a useful predictor if recruitment strength and found that there was enormous variability in the recruitment strength, indicating that other variables likely play a role in determining the production of recruits. Aside from SSB, the authors looked at the usefulness of juvenile relative to larval abundance of fish in predicting recruitment. This test showed that an abundance of later life staged fishes doesn't do a better job in predicting the number of recruits than abundance of earlier life staged fishes. The authors offer a reason for this, saying that these results could have been due to poor quality of the time series collected for older life stages. However, including environmental indicators improved the accuracy of recruitment predictions.

The fourth paper looked at the interactions between predator and prey. What is interesting in marine environments is that often times, the prey species is a predator for the larval stage of its consumer. For example, in the North Sea, adult cods feed on herrings, but herrings also feed on the larvae of cods. As the temperature increased, so did the number of herrings, which caused the population of cods to decrease as well since the larvae were consumed by the large amounts of herrings. Human effects can also exacerbate the decrease in population of the top consumer, so it is important to make sure that fisheries take predator-prey interactions into account in order to avoid shifting the ecosystem balance.

The final paper discusses the population growth rate and certain variables affecting it. Younger populations are common worldwide due to the selection of older fish by fishers since these fish tend to be larger. The juvenation of these populations increases their ability to respond directly to environmental fluctuations. This tells us that interactions between fisheries, environment, and recruitment and population growth have an important and synergistic effect on population dynamics. Durant *et al.*, through their results, showed that shorter generation time causes an increase in recruitment growth, while increasing fishing mortality had a weaker effect. The resilience of the population growth to recruitment change can be attributed to a long-term change in age structure of the population, possibly due to fishing.

Overall, this paper aims to underscore the importance of the connectivity between ecosystem balance and stock of the fish populations. In order to improve recruitment estimations, management strategies must take into account the various types of environmental information describe above. Also, understanding population dynamics of the harvested species most likely has to do with the age-structure of the population, and finally, fish population dynamics can be predicted better by understanding the spatial ecology of the ecosystem.

Lagged Responses to Climate and Area Shifts in Fisheries

Climate changes have had noticeable effects on the global fisheries. Pinsky and Fogarty (2012) focus on the fisheries in northeast United States over the last four decades and examine these warming years to show a northwards shift in fisheries. Sea surface temperatures have increased at 0.23 degrees Celsius each decade there, which is double that of the global average. Surprisingly, fisheries shifted only

10–30% as much as the target species, indicating that economic and regulatory constraints are involved in creating this difference. Anthropogenic factors have played an important role in these changes and as the climate changes, marine fisheries move poleward to accommodate the rise in temperature. Pinsky and Fogarty examine the shifts in selected fish and marine invertebrate distributions and landings over a 40-year period. Fishers in the tropical areas also have to compensate for their lack of fish stock, so they either have to fish more aggressively, or move poleward with the rest of the crowd to keep up with moving fisheries.

Pinsky and Fogarty at Princeton University examine four species of marine animals: lobster, yellowtail flounder, summer flounder, and red hake. These four exhibited poleward shifts in both the spring and fall bottom trawl surveys. They were compared to their target species to see if there was a pattern in the way the species moved. In order to quantify this poleward movement, each species was characterized by its mean latitude, where the species was most commonly found. After knowing the average area at which each species lived, Pinsky and Fogarty used the National Marine Fisheries Service's data on commercial landings (metric tons) and value (dollars) to see a difference in species distribution. Another way they quantified this poleward shift was by calculating the preferred temperature range of species in the landings. Then they compared the mean latitude from the landings against the mean latitude from the surveys by the National Marine Fishery Service to see if there was a poleward shift. They expected northern states to have a higher proportion of total landings as the species moved north and a lower portion as the species moved south.

Results showed a northward shift in all four species' mean latitude and in their landing values. Landings showed much weaker shifts than the target species, however. But, each time a species shifted in latitude, the landed value shifted also. Overall, this study showed that northward shifts in species were mirrored by the northward shifts in fishery landings. States in northern United States also brought in more biomass of fish than before. In addition, warmer-water species were found in most states when average temperatures were warmer.

Although the data seem conclusive, there were some unanswered questions. Firstly, they landings value shifted more slowly than did the target species. Also, since there is a shift in the fishery locations, fishers in tropical areas are unable to catch the same amounts of fish as before. They therefore must be more aggressive in their catching methods. This may lead to even more depletion of fisheries since the migration of fish from south to north gets cut short. However, regulations in the north that have reduced the fishing effort speeds up the shift to the north since most of the population is not being harmed.

Pinsky and Fogarty, therefore, suggest that this shift poleward may not be a bad thing, but instead is a double-edged sword. Some studies predict the loss or severe decline of some fisheries while others predict the growth of fisheries for warm-water species. In the end, clear evidence is presented that changes in species distribution have bottom-up controls at any fishery. However, social and economic factors can cause lags in the way that these fisheries respond to the changes.

Optimizing Databases to Better Fisheries Management Options

Long-term historical data are required to understand ecosystem phenomena such as changes in ecosystem baseline levels, effects of climate change on ecosystems, and trophic cascades. Historical trophic data sets provide time series and spatial data, diet composition and diversity data, and historical food web structures with trophic relationships. However, these things are often hard to find so there is a pressing need to record, spread, distribute and synthesize marine trophic data to understand and deal with massive man-made disturbances. There was a general failure of single species fisheries models and management strategies, which led to an increased call for a multi-species model in the late 80s and later, ecosystem-based management since the 90s. Many reports call for a need to better understand the predators of commercial fish species and their prey items as a part of an ecosystem-based approach to fisheries management. In this study, Simons *et al.* (2013) describe the complicated process in which they will make a Gulf of Mexico Trophic interaction database.

Simons *et al.* report that the sampling locations for food habitats studies had been put directly into the database while ones with a general description were defined by fuzzy polygons. Each study examined a wide variety of fish species with approximately half of the studies examining only a single species. Most of the studies were conducted in estuaries and on the continental shelf and some were located on the shelf slope/deep sea or the mesopelagic realm. Integrating food webs first requires data acquisition, normalization of the data, integration into other data, transformation and analysis or visualization.

Two things are necessary in the architecture of a database: (1) the representation of reality, as in how to most effectively represent concepts or objects in the model, and (2) clear organization of the data so that modeling efforts are made easier. As for geographic and taxonomic boundaries, food habits data include all marine waters in the Gulf as well as estuarine waters. All taxa that inhabit the Gulf for at least part of their life cycles are included, including marine mammals, sea turtles, fishes, sea and shore birds, crustaceans, mollusks, polychaetes, ctenophores, cnidarians, and various parasites. Diets of an organism are largely influenced by the habitat in which it resides and these habitats are classified as essential fish habitats (EFHs). The Costal and Marine Ecological Classification Standard (CMECS) system provides a uniform protocol for identifying, characterizing, and naming ecological units to support activities such as monitoring, protection and restoration. A pilot survey was conducted in order to unify the codes for habitat data in the numerous trophic references using CMECS. There were inconsistencies in the clarity and detail of the data, and in order to alleviate these inconsistencies, all relevant habitat information reported in the document was extracted and adapted to the CMECS terminology.

Metadata records were created for 747 references from a bibliography of food habitats of fisheries for estuarine and marine environments in the Gulf. The database aims to be compatible and accessible for current database projects. The overarching goal is to also be able to have the tools be applicable to large marine ecosystems. The Gulf of Mexico biodiversity database reports 1541 fish species compared to the 762 species in the Gulf trophic interaction database. Some of the species in the trophic interaction database were not included in the biodiversity database because these species are found in brackish and fresh waters located in the upper reaches of estuaries. Included are four classes, forty-four orders, and 236 families of fishes.

An important part of ecoinformatics is extracting information from large ecological databases and then visualizing it. At larger scales, visualization provides a better appreciation for the trophic web's complexity, but often does not create a full understanding of trophic interactions and structure. One issue the authors found was that the database effectively showed the food webs and chains, but in too simplistic a format. They argue that concepts, diagrams, and models that are specific and simple can be used to understand the implications of management decisions on the fate of the species inhabiting the systems of interest to the weekend fisherman or an upper level fishery manager. Another issue the authors address is the scale of the trophic interactions since these interactions have many ramifications and occurrences on different temporal and spatial scales.

Fisheries modeling and management, there has been a large movement towards adopting the ecosystem-based approach but there are few data to implement this system. During a test trial of the model, numerous unexpected results made it evident that significantly more information was needed to fix the database. However, the Gulf of Mexico Trophic interaction database includes both commercially harvested species and other fish species, which allows a more holistic picture of the food webs. The database also has excellent data on food habits of organisms in the Gulf. Eventually, the authors hope that this database for the Gulf will be adopted and modified and applied to large marine ecosystems all over the world.

CAFFEE Predicts Effects of Fisheries Management Options to Mitigate Coral Reefs

Small artisanal fisheries are important to countries in the western Indian Ocean. Developing policy and management strategies involves tactical laws that protect economic, social, and conservation objectives of the nations the fisheries. CAFFEE (coral-algae-fish-fisheries ecosystem energetics) is a model used to evaluate fishing management techniques on coral reefs, developed by Sebastián and

McClanahan (2013). They created models through this program that test the effects of temporary closure on fisheries. In their study, CAFFEE simulated the switching of three types of fishing gears (gill net, spear guns, and hand lines) after severe coral bleaching had occurred, and measured the recovery time of the ecosystem. Overall, researchers found CAFFEE to be useful in the evaluation of management options to mitigate coral bleaching. The model showed that temporary closures from one to three months benefited hand line fisheries more than gill net fisheries. After severe bleaching, gill net fisheries were more apt to recover than were hand line fisheries and when spear guns were used as the main gear in gill net fisheries after bleaching, fish yields became higher.

Sebastián and McClanahan at the Wildlife conservation Society in Bronx, NY and in the University of Cape Town Marine Research Institute in Rondebosch, South Africa developed CAFFEE as an aid to other scientists studying ways to protect these delicate ecosystems. The CAFFEE recreates the coral reef ecosystem with a total of 27 functional groups arranged in trophic levels, like a food web. Light and nutrients are the main energetic drivers of the model and biomass is simulated through energy flow. CAFFEE looks at primary productivity and respiration to get an understanding of the energetics of the system. Different sub-ecosystems are also taken into account, like detrital pools, coral beds, and reef formation and erosion areas. Simulation of coral bleaching includes adjustments of intensity, recovery time, and frequency of bleaching events.

Overall, the study took 40 years and the results were compared to observational series and used to make fine adjustments to the model's internal problems. Three scenarios were examined in this study. The first was the temporary closure of gill net line and hang line fisheries. This part went on for ten years with a fishing effort of five fishermen per km^2. Periodic closures were placed at the end of the annual fishing season. Results showed that these temporary closures had a balancing effect on the catch diversity. This method of management favored recovery of the most vulnerable target fish groups. The gill fishery increased daily catch rates over the ten-year period (12.21 kg per fisherman to 14.6 kg per fisherman), but all closure lengths produced a declining trend in overall total annual catch. The line fisheries had a medium and long-term benefit in annual catch to periodic closures. Five years after closures there was an increase in fish catches due to stock recovery made up for the catch loss caused by closure.

Second, the authors compared the effects of one to three months per year temporary closures in net line and hang fisheries on coral reefs for recovery after severe bleaching. This study went on for fifteen years with a single coral bleaching event of 80% bleaching intensity five years into the simulation and a very slow recovery rate of less than one percent per day. The coral, five months after the bleaching event, decreased cover by three and a half times the original covering. In gill net fisheries, periodic closure had little effect on the coral recovery. In the line fisheries, the same trend was seen. However, the closures had a much larger effect on coral biomass, especially in line fisheries where after three months of closure, the biomass was twice that of non-closure managements.

The final scenario was the switching of fishing gear after a severe bleaching event. The study was also done over a fifteen-year period under the same conditions as the second scenario but the only difference was temporary closures. Instead, fisheries management consisted of maintaining the same gear (gill nets) or switching gears to spear guns or hand lines. The results showed that after bleaching, there was a short-term fisheries yield increase when gill nets were maintained, and when the gear was switched to spear guns. However, when it was switched to hand lines, yields decreased. This was to be expected because after bleaching, macrophyte cover increases, making food more available for herbivorous fish. A substantial portion of the catch of spear guns and gill nets are herbivorous fish, while hand line fisheries primarily catch predatory fish. The third scenario showed that switching from gill nets to spear guns was best for protecting the sea urchin populations. This reduces fishing pressures of the sea urchin predators. In their conclusion, Sebastián and McClanahan say that the acquisition of data for further calibration of their CAFFEE program will require more research.

Conclusions

Marine fisheries are important to keep safe for a variety of reasons and from these articles and we can see that science is making some strides in understanding the complicated dynamics of these ecosystems. MPAs are important to preserving the biodiversity of an ecosystem, but research was not conclusive about how fishers move in response to them. Although they seem to be effective, fishers are a main cause for the depletion so it is critical for fisheries management to understand how they move along with the protected areas. On the other hand, climate change has clearly had an impact on how fisheries have moved. This is important especially for database construction because management strategies have to take into account this trend in increase of global temperature. This could eventually lead to mobile MPAs that move along with the seasons or stricter fishing restrictions. However, it seems critical to be able to understand the patterns of fishers when catching these moving fisheries and plan regulations accordingly.

References Cited

Mesnildrey, L., Gascuel, D., le Pape, O., 2013, Integrating Marine Protected Areas in Fisheries Management Systems: Some Criteria for Ecological Efficiency, Aquatic Living Resources 26, 159–170.

Cariglia N., Wilson, S., K., Graham, N., A., J., Fisher, R., Robinson, J., Aumeeruddy, R., Quatre, R., Polunin, N., V., C., 2013, Sea Cucumbers in the Seychelles: Effects of Marine Protected Areas on High-Value Species, Aquatic Conservation: Marine and Freshwater Ecosystems 23, 418–428.

Horta e Costa, B., Batista, M., I., Gonçalves, L., Erzini, K., Caselle, J., E., Cabral, H., N., Gonçalves, E., J., 2013, Fishers' behavior in Response to the Implementation of a Marine Protected Area, PLoS ONE 8, 1–13.

Cheung, W., W., L., Watson, R., Pauly, D., 2013, Signature of Ocean Warming in Global Fisheries Catch, Nature 497, 365–368.

Durant, J., M., Ottersen, G., Stenseth, N., C., 2013, Impact of Climate and Fisheries on Sub-Arctic Stocks, Marine Ecology Progress Series 480, 199–203.

Pinsky, M., Fogarty, M., 2012, Lagged Social-Ecological Responses to Climate Change and Range Shifts in Fisheries, Climate Change 115, 883–891.

Simons, J. D., Yuan, M., Carollo, C., Vega-Cendejas, M., Shirley, T., Palomares, M., Roopnarie, P., Arenas, L. G. A., Ibañez, A., Holmes, J., Schoonard, C. M., Hertog, R., Reed, D., Poelen, J., 2013, Building a Fisheries Trophic Interaction Database for Management and Modeling Research in the Gulf of Mexico Large Marine Ecosystem. Bulletin of Marine Science 89, 135–160.

Sebastián, C., R., McClanahan, T., R., 2013, Using an Ecosystem Model to Evaluate Fisheries Management Options to Mitigate Coral Bleaching on Western Indian Ocean Coral Reefs, Western Indian Ocean Journal of Marine Science 11, 77

7. Fisheries

Hannah Tannenbaum

The fisheries industry is highly contentious; different stakeholders and values operate on different scales, from national and commercial fishing fleets operating in international waters, to artisanal seasonal fisheries in small island communities, to environmental concerns of stock exploitation as a tragedy of the commons, to considerations of international food security. The overall presiding question in the fisheries remains: how do we ensure the longevity of the industry?

The fisheries industry has been developing and using models to predict the longevity of fish stocks for decades. However, one of the main problems facing the industry is the continued lack of comprehensive or reliable data worldwide. While discrete local fish stocks do exist in regions, overall fisheries longevity is dependent on global fisheries status. Inaccurate reporting of fisheries catches confounds scientific estimates and models on population shifts and dynamics. Burgess *et al.* (2013), Clarke *et al.* (2013), and Pauly *et al.* (2013) all published studies which emphasized the detriments of inconsistent and unreliable fisheries reporting on assessing the viability of the industry.

Besides the paucity of accurate fisheries data, another contemporary theme in fisheries scholarship is the potential effects of global climate change on the longevity of the industry. Global climate change as reflected in changes in ocean temperature reinforces the global quality of the fisheries industry. Fisheries may be affected in a variety of different ways, from range shifts poleward, to changes in phenology based on temperature cues. Mills *et al.* (2013), Heino *et al.* (2013), and Cheung *et al.* (2013) each sought to examine current evidence of how fisheries have responded to contemporary climate change. The question of how stocks will respond to future changes in seasonality and temperature is intimately tied to the above-mentioned limitations of modeling and assessing global population status due to lack of data. Inaccurate fisheries reporting will confound or undermine population shifts due to changing climate.

The last main theme of fisheries scholarship in 2013 was the statistical underrepresentation of developing countries' traditional fisheries. Most catch statistics are based primarily on large-scale commercial fisheries operating in international waters. However, in the developing world, fisheries typically operate on a much smaller, local or seasonal scale, with traditional fisheries playing a key role in both economy, food supply and cultural heritage. Teh *et al.* (2013), and Hall *et al.* (2013) emphasized the importance of traditional fisheries in less developed coastal and island nations both in terms of economic and environmental considerations, but also in terms of food security and cultural significance.

Overall fisheries scholarship is concerned with the sustainability of the industry with good reason. The overall unreliability of fisheries data undermines management efforts to ensure the sustainability. Inconsistent data also make it extremely challenging to extrapolate the future effects of sea surface temperature change on global fisheries dynamics. Lastly, the lack of inclusion of small-scale, traditional fisheries in global statistics and estimates devalues the importance of fisheries as a local industry and as a means of preserving culture.

Modeling Population Dynamics to Prevent Exploitation

Threatened fish stocks have generally only been identified as being overfished and potentially reaching extinction once they have already been exploited and are in population decline. This makes remediation efforts for overfished stocks much more difficult, because much of the damage has already been done. Burgess *et al.* (2013) defined the "eventual threat index" (T), taking into consideration primary effects on overfishing potential. The variables they used to understand exploitation potential include: vulnerability, population size, average catch-per-unit effort, maximum per-capita growth rate, as well as profitability. In particular, this study examined multispecies fisheries, referring to fisheries in which there is a target species, but often the fishery affects many other species as well. Therefore they identified the "key" species as the most commercially valued, or most environmentally sensitive, and therefore the most likely to be exploited. The authors then applied the eventual threat index for historical information on four Pacific tuna and billfish populations, and were successfully able to retroactively predict the eventual declines these fisheries face today. It is intended that this metric can be used to predict the declines of fisheries in the future, before exploitation has occurred.

Many of the most threatened species of fish are caught by multispecies fisheries which catch more species than just the intended catch. Multispecies fisheries are often exploitative because of the fishing techniques used to catch the "key," or commercially desirable species. Burgess *et al* examined historical catch data for four species of tuna and billfish with the eventual threat index they developed, to predict current populations of the species. The eventual threat index for multispecies fisheries was developed based on the assumption that the fates of all species affected can be linked to one "key" species, either the most vulnerable, or the most economically desirable, and therefore the most likely individual species to face exploitation.

Vulnerability of a species was measured using population size, average catch-per-unit effort, and maximum per-capita growth rate. The eventual threat index, T, was calculated by the ratio of the vulnerability of an individual species to the vulnerability of the key species of the fishery in question. Through a series of equations and derivations, the authors were able to generalize the meanings of eventual threat index values: $T>2$ poses a high threat of eventual extinction, $1<T<2$ poses a high threat of overfishing, $0.5<T<1$ poses a possible threat of overfishing, and $T<0.5$ poses a low threat of overfishing. The authors emphasize that changes in technology, equipment, markets and politics can all have an affect on the commercial demand and amount of effort affecting these calculations.

The eventual threat indices were applied to tuna and billfish data from the 1950s to present to see if the current population declines could have been predicted before they occurred. The authors found that T values accurately predicted extinction and overfishing from data as early as the 1950s for marlin and bigeye tuna populations, before the population declines actually started to occur. The purpose of developing the index was to use relatively simple population metrics to predict eventual fisheries declines before they occur, so as to improve overall management while maximizing profits.

From Primary Predator to Picked-on Prey: Shark Fishery in the Pacific Ocean

In 2001 sharks were first listed as endangered species, and since then several measures have been enacted towards their protection. However, the majority of shark fishing is an incidental byproduct of purse seine and long-line fisheries which operate outside of national Exclusive Economic Zones (EEZ). Therefore the effectiveness of international treaties banning shark finning is hard to discern. Another major difficulty in assessing the effectiveness of conservation is the paucity of data on shark population size and structure. Clarke *et al.* (2013) collected and analyzed onboard observer data on shark catches from 1995–2010 in order to evaluate the threat to sharks from commercial fishing, and determine changes in shark populations after finning bans were established. The authors analyzed data on blue, oceanic whitetip, silky, and mako sharks because of their tendency to appear as bycatch in the Pacific tuna fishing industry. Through the analysis of observer data, no clear trend of reduced catches was found consistently for any species, any area, for either type of fishery. The authors suggest that shark retention bans may have a greater impact on population size than finning bans, and that management and monitoring must be made more consistent in order to properly evaluate conservation.

The Western and Central Pacific Fisheries Commission (WCPFC) did not require catch data for sharks until 2011 and observer records are inconsistent and unrepresentative of the whole fishery. Furthermore, there are few data from Chinese, Japanese, and Korean fleets that fish in their own EEZs or in international waters. Clarke *et al.* therefore combined observer data when available with catch-rate analysis to produce standardization models to determine confidence intervals related to catch rate. The authors also used observer data for size-indicator analyses using size as a proxy for age to determine if sharks caught were sexually mature or immature.

Observations were organized into five categories for analysis: retained, finned, discarded, escaped, and unknown. Unknown was removed from analysis. Retained implies the whole shark is kept to be utilized in its entirety versus finned in which the body might be tossed back and just the fin retained. Although the ban on finning was established in 2007, a comparison of before and after finning rates was not appropriate as the various participatory members are currently phasing in the ban gradually.

The results indicated different population and catch rate trends for each of the different fisheries, longline and purse sein, as well as between species of shark. Trends also varied between geographic regions over time. Overall, the oceanic whitetip sharks and blue sharks were found to be in decline over time, despite international efforts to curb shark finning. The authors suggest that the technology is currently not available to completely avoid shark bycatch, and that oceanic whitetip, silky, and mako sharks are more likely to be retained than finned. This evidence clearly shows the undermining of the bans on shark finning, and the authors suggest the ban on finning actually diverts away from efforts to assess population status and conservation. Instead they suggest that in addition to bans on finning and/or retention, there should be additional mandatory monitoring aboard fishing vessels in order to effectively monitor populations with greater precision and accuracy.

Under-reported Overfishing by Chinese Threatens World Fisheries Estimates

Fisheries catch data are the only real means for the fisheries industry, economists, and environmentalists to ascertain the population status of fished stocks. Therefore, accurate reporting of catch data is of the utmost importance. It was discovered in 2001 that China was drastically over reporting their domestic catch in order to achieve the appearance of uninterrupted expansion and success. China has an immense fishing fleet, but is also outside of agreements regarding EEZ and FAO of the UN, and therefore their catch records are important for global estimates, but particularly unreliable as currently reported. While estimates have been made to correct for decades of over-reporting in Chinese domestic fisheries, they are also major participants in distant-water fisheries. Pauly *et al.* (2013) used statistical extrapolations to estimate the Chinese distant-water catches and found severe under-reporting compared to the figures reported to FAO. While the interpolations have high levels of uncertainty, they nonetheless suggest immense inaccuracy of global fisheries catch statistics which has wide implications for employment, economics and ecology.

Recent efforts have been made to improve fisheries reporting through more effective statistical measures for domestic production. However the secrecy of Chinese fisheries reports and the lack of statistical systems independent of the government/fisheries organizations themselves undermines the validity of Chinese reports of distant-fleet catches to the UN FAO. In order to estimate the catch data for China's distant-water fleet, the authors utilized Fermi solutions, inferring unknowns from the known, and a Monte Carlo method to attribute uncertainty values to the estimates.

The steps of this estimation were as follows: 1. Identify nations and years Chinese vessels occupied 2. Record or estimate the number of vessels given local documentation and 3. Assign annual catch estimates by vessel type. These steps were repeated for every nation in whose waters Chinese were found to fish, and then Monte Carlo uncertainty was determined for the overall results. The authors obtained over 500 reports from scientific and non-scientific media sources relating to Chinese vessels in over 93 maritime territories, excluded from only North America, Europe and the Caribbean waters.

It was found that China has severely underreported its distant-water fleet catches. It was estimated that the distant-water annual catch from Chinese vessels is close to 4.6 million t/year whereas the reports to FAO suggest catches of 368,000 t/year. It should be noted that this study was not able to distinguish between legal and illegal catch, as classified as IUU (Illegal, Unreported and Unregulated). Overall the authors suggest it is of the utmost importance that the FAO demand proper reporting and an expansion of fisheries statistics in order for greater accuracy in world food security, economic and environmental estimates.

Mean Temperature of Catch Shows Impact of Ocean Warming on Fisheries

Understanding the impact of climate change on marine fisheries viability has important implications for the sustainability of the industry. Cheung, Watson and Pauly (2013) collected catch and supplemental data, and computed mean temperature of catch, MTC from average inferred temperature of over 900 species of exploited fish weighted with their annual catch rates. MTC was inferred from modeled distributions for the years 1970 to 2006. It was shown that there is a positive relationship between increased rate of SST change and increases to MTC, and that global fisheries have responded with 'tropicalization,' shifts.

Climate change is predicted to have a broad variety of effects on marine fisheries distribution, phenology, and ecological implications including studies which have shown increased vulnerability of some fisheries. Fisheries have been shown to be directly related to optimal physiological tolerances, particularly temperature. Therefore the authors sought to examine the range boundaries and centroids of commercial fisheries over time, and therefore developed the mean temperature catch index to track response to temperature over time.

The MTC was calculated as the average temperature preference of species by their annual catch from 1970 to 2006. The data were further classified by fifty-two large marine ecosystems, LMEs qualified as temperate, subtropical, and tropical. Overall it was found that mean temperature catch increased by a rate of 0.19°C per year, with the largest increases in the northeast Pacific Ocean and northeast Atlantic Ocean. It was also found that there was a close relationship between the rate of SST and MTC changes, suggesting it is a valid proxy for range shifts in relation to climate change.

The relationship between MTC and SST change was undermined by an overall lack of data, as well as increased fishing efforts since the 1970s confounding data irregularities. Nonetheless, the results suggest increased catch rates of warmer water species at higher latitudes and decreases in tropical fisheries. These adaptations to climate change could have important implications on fisheries management and global food security, as many tropical nations are dependent on fisheries for nutritional and economic value.

Selective Pressures Impact on Fisheries Management Reference Points

Fisheries management is a diverse field, essentially reliant on ecological reference points as proxies for fisheries stability and sustainability. Reference points are generally based on population dynamics such as maximum sustainable yield (MSY), spawning stock biomass (SSB), fishing mortality (F), and yield-per-recruit relationships among others. Heino *et al.* (2013) defined fisheries-induced evolution (FIE) as a selective mechanism, which can impact the stock population characteristics and life history traits through the shifting of reference points. Although it has been theorized that fisheries-induced evolution is taking place in various fish stocks, predominantly seen through reduced fish sizes, quantitative evidence for FIE actually occurring in the wild is still unavailable. Nonetheless, the authors conclude that reference points are shifting due to FIE, climate change, or other environmental and ecological factors, and emphasize that reference points must be re-evaluated and adjusted to be current and effective for fisheries management and conservation.

Marine fisheries management uses reference points as a means of establishing targets to both maximize potential catches for food, employment and otherwise, but also to maintain the sustainability of fish populations. Therefore the accuracy of the reference points is paramount to effective fisheries management and prevention of over-exploitation. The authors conducted this study under the overall hypothesis that fishing is a selective pressure that can influence evolution in fish stocks in relation to their development, body size, fecundity and general life histories. They also emphasize that the accuracy or appropriateness of any reference point is limited by the data used to establish it; therefore, the impacts of fisheries-induced evolution must be evaluated to understand how reference points may need to be amended for effective management.

In order to evaluate the impact of FIE, the authors analyze a variety of source material on selective pressures affecting individual-level, population-level, and fishery-level properties. For the individual-level, other research was synthesized to bring the conclusion that FIE leads to earlier maturation, and overall smaller body size, due to fishing increasing mortality of both immature and mature fish, and investment trade-offs between growth and reproduction. Population-level consequences of FIE were less concrete, and again based on the synthesis of others' research. The authors hypothesized that FIE would affect populations in two ways: 1. Populations that are fished can maintain a higher population biomass, or carrying capacity, and 2. If fishing drives stocks towards faster life histories, there will be an initial increase in maximum population growth rate. They came to these hypotheses based on research done on life-history theory and dynamics as well as population demography, which is essentially all theoretical. Lastly, the authors examine the fishery-level impacts due to FIE, namely potentially shifting reference points. Reference points for fisheries management can be based on a variety of different dynamic parameters, including productivity, yield-recruit relationships, or biomass. While reference points can be quantitative values, they are still theoretical measures based on the available data.

Through the analysis of potential FIE effects on the various levels of fisheries dynamics, the authors conclude that FIE is likely shifting reference points. It is emphasized that reference points should not be static quantities, but indeed, must shift and be constantly re-evaluated in order to remain effective and relevant.

Do Fish Sweat? North Atlantic Fisheries Response to 2012 Temperature Anomaly

Sea surface temperature (SST) is a key signaling factor for marine species for a variety of life-cycle events: growth cycles, breeding times, migration times, range location, among others. Fisheries management in the US is cognizant of the importance of temperature, as it has been used as the main parameter for the establishment of species-specific fishing seasons to optimize legal size limit catches annually, as well as the location of set fishing grounds based on seasonal migrations. The authors specifically examined the changes in fisheries after the 2012 Northwestern Atlantic temperature anomaly, in the hopes of extrapolating how fisheries management will be disrupted by changing climate. Mills *et al.* (2013) found that the 2012 temperature anomaly for sea surface was 2°C warmer than the average sea surface temperatures for 1982–2011.

Climate change became tangible for many Americans in 2012 with record high temperatures and the experience of Hurricane Sandy on the eastern seaboard. Temperature changes are important for marine fisheries; temperature norms have allowed for the establishment of fishing seasons for specific seasons, and set fishing grounds seasonally. Mills *et al.* examined the ecological responses to the 2012 temperature anomaly in the North Atlantic. By examining how various fisheries were disrupted spatially and temporally, the authors extrapolated future effects of continued oceanic warming. Ultimately, the authors recommend that fisheries management and regulations, specifically fishing seasons and fishing grounds, must be adapted to climatic changes in order to both preserve the fishing industry and prevent overfishing.

The authors examined the effects of the 2012 abnormal sea surface temperature through the example of the Atlantic lobster fishery. Lobsters migrate inshore annually based predominantly on temperature patterns. Their season for fishing is set for the summer based on the temperatures suitable for moving inshore. The authors found that in 2012 the temperature anomaly affected lobster migration such that there were record catches in June and July, but the season usually peaks in late August. While the higher catch success rate may be advantageous in the short term, warmer temperatures could reduce lobster fitness by cutting short the growing period. Additionally it was noted that the 2012 spike in lobster availability was not actually beneficial but instead led to price crashes for the price of lobster, adding instability to the fishery.

Other fisheries besides the lobster fishery were disrupted or adversely affected by the 2012 sea surface temperature anomaly in a variety of different ways. The disjointed responses to ocean warming will pose a particular challenge for proper maintenance of fishery regulations in the future. The authors suggest that these management institutions must adapt to be more holistic and flexible with their regulatory measures, to take into consideration climatic changes for past regulatory ordinances, and looking towards the future.

Fishery Employment and Fishing for Employment

Marine fisheries are a huge international industry, with a multitude of implications for different sectors: environment, health, employment, trade, industry, cultural heritage, and subsistence. However, accuracy in reporting in fisheries remains a problem, largely due to the lack of consensus of actual worldwide fisheries employment. Teh and Sumaila (2013) quantified worldwide fisheries employment, including direct and indirect employment, and recognition of small-scale fisheries. Using employment and demographic data from 144 nations, as well as the Monte Carlo algorithm to fill in missing data, the authors estimated that some 260 million people are involved in the fishing industry, of which about 22 million are involved in small-scale fisheries.

The accurate evaluation of marine fisheries contribution to worldwide employment may further our understanding of fisheries in terms of socio-economics and the environment. Nations across the world are involved in fisheries but vary in terms of scale and primary purpose. The authors first found that developed countries were typically fishing on a much larger scale than developing countries; this has implications regarding direct and indirect fisheries employment as well as underreporting from "small-scale," artisanal, or subsistence fisheries.

To compensate for the paucity of data regarding small-scale fisheries, the authors first defined small-scale, then classified countries by the UN Human Development Index status and by geographical location, and then used a probability algorithm to estimate the small-scale fishing component. According to the authors, small-scale fisheries operations were classified as "being individuals living in rural areas in the coastal zone." The Monte Carlo algorithm was applied to estimate the number of small-scale fisheries, which uses repeated random sampling with known variables, such as the rural coastal population, or percent below poverty thresholds, to fill in unknown variables, individuals involved in fisheries. The authors recognize that uncertainty remains as to the exact number of small-scale fisheries world-wide, but found that deviations in estimations of small-scale fisheries were most closely matched to coastal population trends. The density and distribution of rural, coastal population was the most important variable in estimating the number of small-scale fisheries in operation; this could provide a jumping off point for future research on fisheries employment.

Ultimately, through national databases and probabilistic estimations, the authors estimated that 260 million people are either employed full-time or part-time directly or indirectly, in marine fisheries. They also estimated that approximately 22 million people were involved in small-scale fisheries, and that 78% of global fisheries workers are from the developing world.

Since global fisheries contribute to national economies, cultural identities, and environmental security, knowledge of the scale and magnitude of the international fisheries industry is crucial to its preservation. The authors were able to use national statistics and probabilities to estimate the world-wide employment of fisheries, and elucidated where further research and reporting is needed.

Sustainable to Whom? Fisheries for Food Security in the Developing World

It is widely accepted that fish stocks worldwide have been over-exploited, and that both commercial and small-scale fisheries face economic challenges due to the lack of certainty in stock security. While studies have focused on the environmental and economic considerations to the fishing industry, few studies have examined the social context of the industry: fisheries as a means of providing national food security. Hall *et al.* (2013) examined the relationship between national GDP (per capita gross domestic product) and dependence on fish for protein in diet, as well as data on wild-caught vs. imported aquaculture for those nations heavily reliant on fish. The authors found that nations that were most dependent on fish as a source of protein and food security were reliant on wild-caught species, and were mostly in the developing world. Ultimately the authors suggest that the site-specific complexity of the international fishing industry demands site-specific, comprehensive management that is inclusive of environmental, economic, and social considerations in policy.

The development of aquaculture has raised expectations that it will surpass wild-caught methods of fishing both in environmental sustainability, not affecting wild stocks, and in terms of economic security. However wild capture fisheries continue to supply the majority of international fish supply as a source of food. In order to examine the relative importance of the fisheries and examine their role as food providers for populations demanding fish for their food security, Hall *et al* collected publically available on production and consumption to draw conclusions.

The authors found that while overall protein consumption increased with national wealth, the proportion of protein from fish was largest for developing countries. Additionally, these developing countries obtained their fish supply predominantly from wild catch fisheries, with small-scale fisheries contributing directly to their nation of origin. It is suggested that for developing countries dependent on fish, particularly African and Small Island Nation States, aquaculture will not replace wild-caught species due to the cost of importation of farmed fish.

Most of the modern scholarship on fisheries is focused on environmental sustainability of the industry, with almost no scholarship preformed on the matter of providing reliable, nutritious food supply for developing nations. This disparity has to do with the extremely diverse and varied nature of the international fishing industry, ranging from commercial factory ships in international waters, to seasonal river fisheries in remote rural locations. International treaties on management are reflective of developed world national values of economics and environment, and not on social relationships and food considerations.

To amend this disparity, the authors posit four broad management goals that can vaguely be applied to most fishing localities: 1. Promote management by those whose values are representative of societal values 2. Promote inclusive stakeholder dialogue 3. Supplement traditional fish catch data with new information including consumer health and values, the 'fish value chain' and 4. Develop intersector support, perspectives and techniques. Additionally the authors suggest a move towards 'sustainability science,' through a re-consideration of the "role of power in shaping the way markets work (and for whom) and the way science informs policy." Ultimately, the authors argue that overall fisheries management must be improved in order to ensure the environmental, economic, and social sustainability of worldwide fish production and consumption.

Conclusions

The sustainability of fisheries has implications globally. As an industry, fisheries contributes to the global economy in terms of trade, and employment, to ocean biodiversity and security, to international politics through licenses, and to cultural heritage through traditional fisheries. Each of these facets will be affected by fisheries-shifts due to global climate change, and to further global stock declines. The biggest challenge facing the international fisheries industry is the overall unreliability of fisheries data. Until there is a comprehensive global system of fisheries statistics and reporting, studies evaluating current and future fisheries status cannot draw conclusions to inform management. It is paramount for international fisheries management that fisheries reporting be overhauled.

References Cited

Burgess, M *et al.* 2013. Predicting Overfishing and Extinction Threats in Multi-species Fisheries. PNAS early edition, 1–6.

Cheung, W, Watson, R and Pauly, D. 2013. Signature of Ocean Warming in Global Fisheries Catch. Nature, 497, 365-369 doi: 10/1038/nature12156

Clarke, S *et al.* 2013. Population Trends in Pacific Ocean Sharks and the Utility of Regulations on Shark Finning, Convention on International Trade in Endangered Species of Wild Fauna and Flora CITES, Sixteenth meeting of the Conference of the Parties, Bangkok (Thailand), 3-14 March 2013

Hall, S. Hilborn, R. Andrew, N. *et al.* (2013) Innovations in Capture Fisheries are an Imperative for Nutritional Security in the Developing World. PNAS 110, 8393-8398.

Heino, M. *et al.* 2013. Can Fisheries-induced Evolution shift Reference Points for Fisheries Management? ICES Journal of Marine Science 70, 707–721.

Mills, K. et al. 2013. Fisheries management in a changing climate: lessons from the 2012 ocean heat wave in the Northwest Atlantic. Oceanography 26, 1–6.

Pauly, D. Belhabib, D. Blomeyer, R *et al.*2013 China's distant-water fisheries in the 21st century. Fish and Fisheries. DOI: 10.1111/faf.12032

Teh, L., Sumaila, U R. 2013. Contribution of marine fisheries to worldwide employment. Fish and Fisheries 14, 77–88.

8. Acidification, Warming and Coral Reef Ecosystems

Dawn Barlow

Global climate change is known to have severe impacts on marine ecosystems. Coral reefs are among the most intricate and biodiverse ecosystems on earth, and are also arguably among the most delicate and susceptible to climate change. As the amount of CO_2 in the atmosphere increases, global temperatures rise, causing sea surface temperatures to rise as well. Additionally, the oceans act as sink and absorb CO_2 from the atmosphere, and this dissolved carbon dioxide causes the oceans to become more acidic, lowering the pH of seawater. While perhaps the most severe and ongoing damage to coral reefs originates from atmospheric CO_2 levels, there are many local stressors that contribute to their degradation. These could be runoff from agriculture and development, overfishing that causes an imbalance in the trophic levels, disease, boat traffic that causes physical breakages in the coral from anchoring, and many other sources of stress.

Many researchers have been looking to address the current state of the reefs and predict how they will respond under future ocean acidification and warming conditions. Because of the complexity of coral reef ecosystems it is difficult to predict just how dramatically they will respond to projected future conditions, and so there are many studies that analyze just one piece of the system at a time or just one factor that influences corals' overall health. This chapter is an attempt to bring together several of the pieces of these complex systems that have been studied in great detail to provide a more holistic overview of what, according to the scientific literature, the current standings and predictions are for the health and integrity of coral reefs.

The structural framework that provides a home for all the life that comprises a coral reef ecosystem is made of the calcium carbonate skeletons of calcifying organisms. Calcifying organisms such as corals and encrusting algae use carbonate ions in the seawater to form their skeletal structure. As the oceans acidify, there is a decrease in the overall concentration of carbonate ions and increase in the concentration of bicarbonate ions. As it turns out, corals do have some ability to utilize bicarbonate to form their skeletal structure, though not as well as carbonate. Corals maintain and regulate an internal pH at the tissue-skeleton interface, where calcification actually takes place. In this way they are also able to buffer their response to acidification to a certain extent, until the external seawater pH drops below a threshold point after which they are unable to continue to calcify.

These examples of how corals may be able to mitigate their responses to increasing acidity examine the direct effects that lowered pH will have on coral health. However, corals are part of a dynamic ecosystem, and so there is a whole host of indirect effects that impact the corals' resilience to acidifying waters. Many algae tend to flourish in lower pH conditions. Microalgae have an ability to form mats that compete with corals for space, causing a phase shift in ecosystem dominance. The weakened ability of corals to calcify is compounded by the fact that they are being inhibited by mat-forming algae which not only outcompete the corals for space but also impair larval recruitment. Herbivorous fishes and other grazers such as urchins can often play a role in keeping the algae at manageable levels for the corals, however those species are also susceptible to overfishing and disease, and the role that they play is often limited in these scenarios.

Microborers, photosynthetic microalgae that erode coral structure, also tend to do better in lower pH conditions as well as at warmer temperatures. These microborers cause the dissolution of coral skeletons, and it has been found that the rate of calcification of corals is the slowest in scenarios where both acidity and temperature are elevated.

While the detrimental effects of increased seawater acidity can often be buffered by the corals, raised seawater temperatures have been shown to have significant and deleterious effects on reef ecosystems and it is far more difficult for the corals to protect themselves against warming waters than against lowered pH. As far as the early life stages of coral are concerned, larvae do not appear to be severely influenced by acidic waters but their rate of development and therefore population connectivity are significantly impacted by raised temperature conditions alone.

In a study of an isolated reef system far removed from human impact, roughly 90% of shallow water corals were wiped out when the water temperature increased significantly in a natural sea-surface temperature warming event and was maintained at an abnormally high temperature for a two-month period. In other, less isolated reefs, larval recruits could potentially come from elsewhere and repopulate the areas of dead coral. But the less isolated reefs were also under greater amounts to stress due to human activity and overfishing, and so even though the isolated reef could not recruit from other reefs in order to repopulate. The presence of abundant herbivorous fishes, resulting from the absence of human activity, meant that the reef could be completely reestablished within a decade of the warming event. This is an example of a case where increased temperatures were maintained for only a short period and the reef was able to recover, however it does not indicate that corals will have the ability to acclimatize if those temperatures would be maintained indefinitely.

This issue of acclimatization to warmer as well as colder temperatures has also been studied by examining how coral colonies behave when they are transplanted to regions whose temperatures exceed the historical thermal regimes to which they have grown accustomed evolutionarily. Evidence points toward genetic make-up being what will ultimately control the ability of corals to acclimatize, because once the thermal range of the corals has been exceeded, any more thermal tolerance requires adaptation through selection of tolerant genetic variants.

Many of the studies referenced examine what the fate of the coral reefs will be if acidification and warming trends continue into the future, and these projections are based on the continuation of current human practices. However, if there is any hope of slowing the degradation of the reefs, dramatic policy measures will need to be put into place. There are studies that model what the future health of the reefs could look like if protection measures are implemented for the short-term, such as local fishing restrictions and protection of herbivorous grazers, as well as for the long-term, such as global reductions in CO_2 emissions. What such studies have found is that although local management and protection will have some definite preventative effect on the degradation of reefs, dramatic global CO_2 emissions reduction measures will be necessary if we wish to see lasting change.

Corals May Use both Carbonate and Bicarbonate Ions for Calcification in Response to Ocean Acidification

The highest known marine biodiversity is hosted by the calcium carbonate framework of corals and calcifying algae that form rich and dynamic coral reefs. Coral reefs are threatened by ocean acidification caused by an increase in dissolved carbon dioxide in the water, and this in turn decreases carbonate ion concentrations and increases bicarbonate ion concentrations. Comeau *et al.* (2013) evaluated the roles of carbonate and bicarbonate in the calcification of coral and crustose coralline algae in both light and dark conditions. They found that coral can maintain present-day calcification rates in the light if the decrease in carbonate concentrations is compensated by an increase in bicarbonate concentration, but this was not sufficient for calcification in the dark. Though crustose coralline algae were able to calcify using bicarbonate, it was not sufficient to compensate for the decrease in carbonate that is a result of ocean acidification. The results of this study show that the response of tropical coral reef communities to ocean acidification might be less dramatic than previously predicted due to the ability of calcifying organisms to utilize bicarbonate while it is light, but the decrease in the ability to calcify in the dark will undoubtedly result in an overall reduction in the calcifying ability of coral reefs in increasingly acidic oceans.

This experiment conducted by Comeau *et al.* was unique in that instead of simply increasing the acidity of the water that the calcifying organisms were in, they addressed the effects of varying concentrations of carbonate and bicarbonate ions. The carbonate chemistry of the water used in this lab experiment was manipulated using carbon dioxide-equilibrated air, and varying carbon dioxide treatments were created by bubbling ambient air, carbon dioxide-enriched air, and carbon dioxide-depleted air into tanks of seawater. One coral species, *Porites rus*, and one crustose coralline algae species, *Hydrolithon onkodes*, were used as study species of calcifying organisms for this experiment. Buoyant weight measurements were used to evaluate calcification over the two-week period in which this study was conducted, and the alkalinity anomaly technique was used to examine short-term calcification in light and dark conditions.

The methods used by Comeau *et al.* allowed them to examine the relative concentrations of carbonate and bicarbonate and the calcifying ability of coral and crustose coralline algae in both light and dark. They found that for coral, calcification in the light was affected by both carbonate and bicarbonate concentrations, and in the dark by carbonate but not bicarbonate. For crustose coralline algae, light and dark calcification were affected by both carbonate and bicarbonate concentrations. The implications of the fact that the coral was better able to calcify in the presence of bicarbonate in the light are consistent with other studies, which couple calcification with photosynthesis and demonstrate that photosynthesis can, under some conditions, be stimulated by additions of bicarbonate. Carbonate is important for calcification, and bicarbonate is important for both calcification and photosynthesis. However, the way that ocean acidification works is that increasing bicarbonate means decreasing carbonate, and so the fact that bicarbonate increases calcification in the light is offset by the fact that overall calcification is decreasing due to the declining amount of carbonate ions in the water.

These results provide more insight into the rate at which reef-building organisms can calcify in increasingly acidic ocean waters. The fact that there appears to be a positive correlation between bicarbonate and calcification in the light shows that the future ability of coral reefs to calcify may not be quite as dire as previously thought. But ultimately the decreasing concentrations of carbonate that come with increased acidity mean that the increased ability to calcify in the light will be offset and the reef-building organisms will struggle to calcify as ocean acidification continues to be an issue in the future.

Seawater Acidity and the Physiological Ability of Corals to Calcify

This study by Venn *et al.* (2013) addresses how ocean acidification reduces the calcification rate of corals by reducing internal pH at the calcifying tissue-skeleton interface. They aim to predict how corals will respond and potentially acclimate to ocean acidification by looking at how acidification impacts the physiological mechanisms that drive calcification itself. Coral skeletons are formed from calcium carbonate crystals (aragonite), produced in the fluid-filled subcalcioblastic medium (SCM), which underlies the calcifying tissue. The calcifying tissue elevates pH the SCM relative to the pH of the exterior seawater, favoring the conversion of bicarbonate to carbonate, and enhancing precipitation at the site of calcification. This ability of corals to regulate internal pH is anticipated to be critical in their resilience to ocean acidification, and overall, findings from this study suggest that reef corals may be able to mitigate the effects of seawater acidification by regulating pH in the SCM.

This study is an *in vivo* investigation of whether CO_2-driven acidification of seawater causes changes in pH at the tissue-skeleton interface and rates of coral calcification. Venn *et al.* performed all of their experiments on 1-cm^2 microcolonies of the coral *Stylophora pistillata*. The carbonate chemistry of the seawater in three of the four tanks used was manipulated by bubbling CO_2 to reduce pH and create different levels of seawater acidity to work with. The small surface area of the samples was chosen to achieve stable pH and carbonate chemistry. Live tissue imaging of the samples was done in a perfusion chamber, where oxygen levels were maintained at a stable level. The effects of seawater acidification were measured over both short and long timescales. Calcification was measured using confocal microscopy to obtain very fine-scale measurements of the aragonite crystals themselves.

Venn *et al.* found that for corals that had been exposed to seawater with reduced pH for more than a year, the pH in the SCM was significantly lowered. The pH in the calcifying tissue was significantly lowered in the scenario with the most reduced seawater pH as well. Results were very similar on a much shorter timescale of only 24 hours, in which the authors found significant decreases in the pH of the SCM in the two most reduced seawater pH treatments. Statistical analysis showed no significant interaction between treatment and time. In the calcification experiment, the authors found that aragonite crystals grew significantly less in the most reduced pH treatment. The same was true when calcification was examined on a colony scale, where the corals exposed to the most acidic seawater showed a significant decrease in growth.

How calcifying organisms such as corals will respond to increasingly acidic seawater conditions will depend on their physiological ability to regulate internal pH in the fluid at the tissue level and in the calcifying cells. In this study, internal pH and calcification were both reduced significantly only at very low external seawater pH, which indicates that *S. pistillata* may have a relatively high tolerance to changes in seawater chemistry though they are susceptible to reduced calcifying ability in extremely acidic conditions. The calcification experiment showed agreement between the growth of individual crystals and colonies, showing significant reductions in growth at the lowest pH treatment, demonstrating a physiological tipping point after which rates of calcification rates significantly decline. This study investigates the physiological mechanisms of coral calcification in a controlled laboratory setting. The authors suggest that valuable future investigations might include similar organism-scale investigations, as well as field investigations that take into account variables that are not accounted for in the laboratory.

Dissolved Carbon Dioxide Acts as a Resource for Mat-Forming Algae to Outcompete Corals and Kelp Forests for Space

This study addresses the effects of enhanced CO_2 levels in the ocean by looking at how increased acidity might indirectly cause phase shifts in community structure of coral reef and kelp forest ecosystems in temperate and tropical waters. Under elevated acidity and temperature conditions, productivity of certain photosynthetic organisms such as mat-forming algae (low-profile ground-covering macroalgal and turf communities) can increase, making CO_2 not only a direct stressor but also an indirect stressor by being a resource for certain competitive organisms, creating enormous potential for shifts in species dominance. Additionally, ocean acidification acts together with other environmental stressors and primary consumers, and these factors also influence community response to acidic conditions. Connell *et al.* (2013) investigate the prevalence of mat-forming algae in three different scenarios where CO_2 levels were either ambient or elevated: in the laboratory, in mesocosms in the field, and at naturally occurring CO_2 vents that locally alter the seawater chemistry. They find that in all the scenarios, the algae mats respond positively to the elevated conditions, increasing growth rate and cover so that the algae became a majority space holder regardless of any herbivory. This is likely because the new environmental conditions favor species with fast growth and colonization rates and short generation times, and these are the species that are capable of completely displacing space-holding species with slower growth and longer lifespans.

In this study, Connell *et al.* attempt to examine the indirect effects that ocean acidification has on coral reef and kelp forest ecosystems due to competition for space under elevated CO_2 and temperature conditions. This is a very broad investigation, where several different techniques are carried out in several different locations. All of the acidification questions are addressed for both temperate and tropical regions—the temperate experiments are carried out in the Mediterranean and the tropical experiments are conducted on the Great Barrier Reef. One thing to keep in mind when considering this study is the fact that that these individual investigations in different regions were carried out over different timespans and using different techniques in addition to being conducted in different places. The idea behind using these three different techniques for each region is to get the most holistic picture of how entire communities will respond to elevated conditions. In the laboratory it is possible to easily eliminate certain variables that could be interfering with the results, but on the other hand it is difficult to account for all the factors in an ecological community in a laboratory setting. The mesocosms allow for an experiment to be carried out in the field but with certain manipulated factors, in this case pH. The future pH conditions were chosen to represent projected near-future conditions anticipated for the end of the century; 7.7 –7.8 in the temperate region and 6.8 –7.8 in the tropics. The vents allowed for an observational study on the existing community under naturally elevated CO_2 conditions where local ecological communities have adapted. The drawback with the field studies at these vents is that the spatial and temporal variation of pH in these acclimatized communities does not necessarily behave as would future ocean conditions, although the fact that it is an observational study eliminates the possibility that manipulations are skewing the results.

Connell *et al.* found that in all cases, mat-forming algae cover increased under elevated CO_2 conditions, regardless of herbivory. Net productivity in the lab increased significantly, but there were several sources of potential loss that were not accounted for in the laboratory. In field mesocosms, the growth rate of the algae mats increased under elevated conditions to two to three times what it is under ambient conditions. At temperate vents, the algae mats covered 40% more space and at tropical vents, 50% more, in both cases expanding from a minority space holder to a majority space holder. In the community phase-shift that took place, calcareous taxa (primarily crustose coralline algae and barnacles) maintained a percentage cover in the elevated conditions that was similar to the ambient one in the early stages of succession, but were then quickly overgrown by mat-forming algae after 3.5 months. Mat-forming algae appear to inhibit the percentage cover of other taxa as well. Whether there was a change herbivory was not made clear by this study, as densities of calcareous grazers such as urchins did not differ between ambient and elevated conditions, and the densities of herbivorous fish at the vents was is unknown and fish had the ability to easily move in and out of the low pH area. This suggests that the increase in cover of algae mats is not due to an absence of grazers, and that they still do hold an important role in mediating competitive interactions between space holders.

This study demonstrated the fact that under projected future acidification conditions, species with fast rates of colonization and growth and short generation times are favored and competitively displace slower-growing and longer-living space holders. This shows that the negative effect on coral growth rates that happens in the presence of increased CO_2 is strengthened by increased growth rates of mat-forming algae. The authors of this study suggest that because CO_2 acts as a non-additive stressor in combination with local stressors, the potential for acidification-related phase shifts in response to gradual acidification is likely to increase with additional stressors. The implications of this finding are that by reducing local stressors such as nutrient pollution and overfishing, it is possible to reduce the effects of global stressors such as ocean acidification and temperature, and that reduction of these local stressors is something that needs to happen for effective management if there is hope for the reduction of loss of coral reef and kelp forest ecosystems.

Erosion of Corals by Photosynthetic Microborers under Increasing Ocean Acidification and Warming

Microbioerosion in corals is a process that is caused by chemical dissolution and driven by metabolic activities of internal microborers within coral skeletons. This study hypothesizes that the increase in temperature and acidity of seawater that is projected to take place in the coming decades will reduce calcification as well as increase dissolution in corals and crustose coralline algae because it will alter microbioerosion processes on coral skeletons. Reyes-Nivia *et al.* (2013) compared current ocean conditions with two elevated CO_2-temperature scenarios to explore how skeletons of branching and encrusting corals respond combined acidification and warming, and to see if these elevated scenarios would alter the microbioerosion processes. They measured the rate of calcification under these varying conditions, and found that dissolution of coral skeletons was driven predominantly by photosynthetic endolithic algae microborers, and that this was increased with combined acidity and warming. They also found that that the projected acidification and warming scenarios appear to favor the accumulation of endolithic algae, which would lead to significant bioerosion within coral reef frameworks.

Reyes-Nivia *et al.* conducted this experiment in the lab using a system of custom software that allowed for CO_2 and temperature to follow seasonally appropriate fluctuations measured at a reference field site, providing a near-perfect match to the control conditions. In the lab, three different acidity-temperature scenarios were created by bubbling different amounts of CO_2 into tanks that were maintained at different temperatures using industrial heater-chillers. As the aim of this study was to examine the activity and abundance of microborers on coral skeletons in these different scenarios, samples of a branching (*Porites cylindrica*) and an encrusting (*Isopora cuneata*) coral harboring high amounts of endolithic algae were collected and the coral tissue was removed from the skeletons to simulate recently dead coral substrates before they were placed in the different tanks. A subset of these samples was kept in the dark so as to inhibit the reestablishment of these endolithic algae which require sunlight to thrive. The pH was measured in the water as well as within the coral skeleton through drilled holes throughout the study. Total calcification and bioerosion was measured as change in the amount of calcium carbonate using the buoyant weight method, which was also used to measure monthly microbioerosion rates normalized to the total surface area. The loss-on-ignition method was used to calculate the biomass of the endolithic algae at the end of the duration of the experiment.

What Reyes-Nivia *et al.* found was that, under natural day-night cycles, all of the coral skeletons with endolithic microalgae lost calcium carbonate. Under elevated CO_2-temperature conditions, the total microbioerosion increased in both types of coral. For the coral skeletons under full dark conditions—those without photosynthetic endolithic algae—there was no sign of dissolution; in fact there was a positive increase in buoyant weight.

This study showed that the dissolution of coral skeletons, which is driven predominantly by photosynthetic microborers, increased under combined ocean acidification-warming scenarios, and that this dissolution varied with the specific scenario and the coral substrate. This variance shows that the skeletal structure of the coral—microskeletal architecture, porosity, density, minerology—plays an important role in determining the magnitude of how the acidity-warming scenario will affect coral skeletal dissolution. The projected acidification and warming scenarios appear to heavily favor biomass accumulation of photosynthetic microalgae, which in turn would mean significant bioerosion of coral reef frameworks.

The implications of this study are that the conditions within projected future oceans appear to influence both the biological and ecological responses of endolithic microborers, which lead to the dissolution of coral skeletons. This is demonstrated by the enhanced biomass, shifts in community structure, and increased respiration rates of endolithic algae under elevated CO_2-temperature conditions. This implies that if the oceans continue to acidify as predicted, not only will the corals not be able to calcify as well in more acidic waters, but the increased abundance of photosynthetic endolithic algae will contribute simultaneously to their dissolution.

Comparing the Near-Future Effect of Temperature and Acidification on Early Life History Stages of Corals

Both ocean temperatures and pH are projected to increase due to climate change in the near future—it is predicted that temperatures will be raised by 2°C and that acidity will increase by ~0.2 pH units by the end of the century. While much investigation has been done on the effects of temperature and acidity on the ability of adult corals to form the structure necessary to maintain the integrity of the reef, Chua *et al.* (2013) investigated the direct effects of increased temperature and acidity on the early life history stages of corals. They looked at fertilization, development, survivorship, and metamorphosis of coral larvae under control conditions as well as under elevated temperature and acidity, both separately and in combination. When the two factors were combined, the results were inconsistent. Overall, the conclusion drawn from this study was that acidification alone is unlikely to be a direct threat to early life history stages of corals, at least in the near future. Increasing temperature, on the other hand, was found to increase the rate of larval development and thereby affect coral population dynamics by changing patterns of connectivity.

Chua *et al.* carried out this investigation in the lab using larvae samples of *Acropora millepora* and *A. tenius* collected from the Great Barrier Reef and kept in aquaria until spawning. The larvae were exposed to four treatments: ambient temperature and ambient pCO_2, ambient temperature and elevated pCO_2, elevated temperature and ambient pCO_2, and elevated temperature and elevated pCO_2. Experimental temperatures were maintained using aquarium heaters, and experimental pCO_2 levels were maintained using a CO_2 mixing system developed by Munday *et al.* (2009) that involves bubbling mixed gasses through sump tanks. The two levels of CO_2

concentration and temperature were selected based on the European Project on Ocean Acidification protocol.

The results of this study were highly variable, and given the number of treatments tested, not many conclusions could be drawn from the results. No interactions were shown between temperature and pCO_2, but during development, motility was achieved more rapidly by the coral larvae under elevated temperature conditions. The effects of elevated temperature and pCO_2 were also variable between the species of coral and treatments, but overall mean survivorship trended toward being lower under combined elevated conditions. The strongest result was that temperature on its own accelerated development. Elevated pCO_2 did not have any effect on metamorphosis, either on its own or in combination with elevated temperature. Neither temperature, elevated pCO_2, nor the combination of the two had any significant effect on fertilization.

From the results of this experiment, Chua *et al.* concluded that elevated pCO_2 rarely affects development in early life stages of corals, and only in combination with elevated temperature, contrary to previous predictions. When temperature was increased by 2°C the early life stages of development of corals were affected, which is consistent with metabolic theory. Though some coral species may be more susceptible to temperature and elevated pCO_2 stress than others, the overall conclusion of this study is that acidification will likely have a minimal effect on the ecology of early life history stages of corals. The ecological implications of more rapid rates of development under elevated temperatures are likely to be highly dependent on the region, species, and local conditions such as reef density and hydrodynamics. In the near future, the temperature increase rather than the increase in acidity associated with global warming is more likely to have ecological consequences for the early life stages of corals.

Coral Reefs have the Ability to Recover from Severe Disturbance in the Absence of Human-Induced Stressors

This study looks at the potential for isolated coral reef ecosystems to recover after severe disturbance, in this case ocean warming events. The study system that is examined by Gilmour *et al.* (2013) is Scott Reef, an oceanic reef at the edge of Western Australia's continental shelf. Because Scott reef is so far removed from other reefs, it means that there is a lack of connectivity that makes it far more susceptible to disturbances because recruits cannot easily come from elsewhere. Additionally, this isolation means that there is negligible human activity that could stress the system and inhibit its recovery. In 1998 there was a catastrophic bleaching event caused by a sudden extreme warming in sea surface temperatures, which led to a mass coral die-off in regions across the Pacific. It was not expected that the Scott Reef system would be able to return to pre-disturbance conditions, and if it did recover the projections were that it would be decades before it reached anything comparable because all recruits would have to be locally produced—none could come from outside. However, since the area is not fished there are large populations of herbivorous fish to keep down the algae that would otherwise take over a disturbed reef and inhibit recruitment. Within twelve years of the bleaching event, Scott Reef was able to recover to a point where reproductive output and recruitment were comparable to pre-disturbance levels.

In this study, Gilmour *et al.* did not manipulate pieces of the ecosystem. Instead, they used an existing system for which they had data for coral abundance and recruitment rates as well as ocean temperatures, and then monitored these same things during and after the mass bleaching event. They surveyed permanent transect lines on the reef, which they categorized as either reef crest (~3m), reef slope (~9m), and upper reef slope (~6m), depending on the depth of the coral. Recruitment was monitored using settlement tiles, which were deployed at regular intervals along the transect line. Thus they were able to observe the correlation of these factors and influences as the reef recovered after severe disturbance.

In February of 1998, seawater temperatures at Scott Reef rose to an unheard of 13.3°C and remained at extreme high temperatures for the next two months. In the next six months, 80–90% of live coral cover was lost from the reef crest and reef slope, and nearly 70% was lost from the upper reef slope. Recruit numbers decreased to zero from the previous years' 2600–5600, and in the following six years recruitment rates were less than 6% of years prior to the disturbance event.

In other systems that have been studied following severe disturbance, what usually takes place is a phase shift to macroalgae which outcompetes the corals so that there is no longer a substrate on which recruits can settle. But what happened at Scott Reef was that the substrata that were made available by the dead corals were colonized by fine turfing and crustose coralline algae. There was also an observed increase in the densities of herbivorous fishes following the event, and the inability of macroalgae to colonize the dead coral was very likely due to the dramatic increase in grazing capacity by large herbivorous fishes. This surplus in grazing capacity in the Scott Reef system was able to assist subsequent coral recruitment, allowing for more coral to be locally produced as well as giving them a place to settle and therefore increasing overall survival. On reefs experiencing chronic pressures from human impact, the mean survival of recruits tends to average around 50%. On Scott Reef during this recovery period however, mean survival of recruits ranged from 83–93%. Within a decade of the bleaching, reproductive output and recruitment were similar to the levels prior to the bleaching, and after 12 years the overall health of the reef was determined to be comparable to the time before the disturbance.

This study shows that anthropogenic pressures, particularly fishing, are dramatically inhibiting coral reef recovery after disturbance if a system as isolated as Scott Reef was able to recover from 90% coral loss without recruits from other populations within a decade, simply because of the absence of human activity. This observational study demonstrated the importance of having an intact ecosystem—the fact that the herbivorous fish had not all been caught—and the importance of limiting anthropogenic stressors if we aim to protect the reefs from further degradation. The authors of this study suggest that creating marine protected areas could be a key to promoting coral reef resilience through protecting healthy, intact ecosystems.

Thermal Regimes define both Cold and Warm Temperatures that limit Coral Acclimatization

Over just the past few decades ocean temperature has contributed to significant losses in global coral cover, and the extent to which corals can undergo physiological acclimatization or genetic adaptation to thermal changes remains uncertain. However, this information will be crucial for the effectiveness of conservation strategies and accuracy of projections of reef futures. This study conducted by Howells *et al.* (2013) investigates the potential for corals to acclimatize to temperatures that exceed historical thermal regimes. This is done by investigating several parameters—bleaching, mortality, *Symbiodinium* type fidelity, and reproductive timing—in coral colonies that have been transplanted between warm central regions and cool southern regions of the Great Barrier Reef for a period of 14 months. This particular study investigates the thermal tolerance of both the coral hosts as well as their symbiotic partners, dinoflagellates known as *Symbiodinium*. Bleaching of corals takes place when the capacity of the photoprotective and antioxidant mechanisms in *Symbiodinium* cells to prevent and remove damaging reactive oxygen molecules from leaking into the coral tissue is exceeded, and dissociation takes place between the coral and *Symbiodinium* caused by oxidative stress. Both coral and photosynthetic *Symbiodinium* acclimatize seasonally to changing temperatures using various mechanisms. But this study discusses the fact that acclimatization is ultimately limited by the genetic make-up of both partners in the symbiosis, because once the capacity for acclimatization is exceeded, any more thermal tolerance requires adaptation through selection of tolerant genetic variants. The authors found that there was significant bleaching and mortality in all of the transplanted corals, in both regions. Additionally, there was a shift of *Symbiodinium* types in the transplanted coral in the warmer region, but this did not prove to be more photosynthetically efficient.

Howells *et al.* collected and reciprocally transplanted fragments of coral colonies between two inshore reef sites in the central and southern regions of the Great Barrier Reef. The study species of coral used for this study was *Acropora millepora*, as it is common and sensitive to bleaching. *Symbiodinium* types hosted by the different *A. millepora* populations differed between the two sites, with heat-tolerant *Symbiodinium* type D at the central site and heat-sensitive *Symbiodinium* type C2 at the southern site. The coral fragments were transplanted at the beginning of autumn and monitored at 1- to 4-month intervals over a period of 14 months, when coral condition, *Symbiodinium* type, photochemical efficiency, reproductive timing, and growth rate were all recorded.

At the southern site, the coral fragments transplanted from the central site remained healthy as temperatures began to decrease for the first several months while staying in the thermal range of the central region and even as it grew colder than the typical winter minimum. However, after the main cold period in the year a cold-water bleaching was observed, where 50% of the transplanted coral fragments suffered partial- or whole-colony mortality, and the coral fragments from the central region that were still alive had bleached. Bleaching was sustained even after seawater temperatures began to warm again. At the central region, the transplanted corals remained healthy as the seawater temperatures cooled. But then when temperatures began to rise again and exceeded the typical summer maximum for the southern region, the transplanted corals all bleached, and shortly after that 50% were dead and those that were alive had very little surviving tissue. The authors found that for the most part the *Symbiodinium* type remained constant in coral fragments for the duration of the study. However, in the southern corals that were transplanted to the central region, they observed that after bleaching there was a shift from *Symbiodinium* type C2 to type D. There was no difference in photosynthetic efficiency between the types—pigment was equally low in all bleached fragments whether they were associated with type D, C2, or a mix of the two. For both coral genotype transplants, coral growth was substantially slower regardless of whether or not the fragments showed visual signs of temperature stress. Some of the transplanted corals spawned up to a month earlier than their counterparts at their native sites.

Howells *et al.*'s reciprocal-transplant study demonstrated the poor ability *A. millepora-Symbiodinium* symbioses to acclimatize to temperature changes that exceeded the upper and lower limits of their local thermal regimes. This shows that the thermal tolerance of this particular species is likely determined primarily by genetic adaptations to local thermal regimes.

The *Symbiodinium* types associated with the corals in the different regions studied probably had a strong influence on the thermal tolerance of the corals. Though there was a possibility that corals could withstand bleaching by changing *Symbiodinium* partners, at least in the short term, results from this study showed that changing *Symbiodinium* partners was not sufficient in enabling the host corals to cope with prolonged and extreme changes in temperature.

While these results highlight and confirm the importance of conserving coral populations, they also demonstrate that coral restoration through transplants is unlikely to be successful if the temperatures of the donor and recipient reefs are not correctly matched up. While it had been previously thought that it could be feasible to transplant corals from warmer regions to colder waters, this study demonstrates that corals also have a lower thermal limit. The authors suggest that future research is needed to investigate whether success of coral transplantation could be improved by using individual samples from multiple donor sites across ecological and thermal gradients.

Conservation Measures and using the Fates of Past Reefs to project Future Scenarios

Kennedy *et al.* (2013) examine the structural and ecological human-caused damage on coral reefs in the Caribbean since the 1960s and then generate modeled predictions for what coral reefs in the Caribbean might look like between now and the year 2080 if practices of local conservation and global action were to be implemented. They examine driving ecological factors of Caribbean reefs, reconstructing past disturbances to address the important roles that all of the many pieces play in contributing to overall health of the reefs. Some of these disturbances between the 1960s and the 2000s include depletion of the reef due to overfishing, loss of branching corals because of disease, bleaching, and bioerosion, hyperabundance of urchins because their predators were lost to overfishing, loss of urchins because of disease, poor watershed management that has led to changes in nutrient abundance, and ongoing climate change since the 1960s. In order for a coral reef to maintain its structure, the rate at which carbonate is produced must be greater than the rate of erosion—the carbonate budget must be positive. In order to maintain a carbonate budget that is either positive or at equilibrium, Kennedy *et al.* suggest that local management for the protection of grazers such as parrotfishes is important and can create a positive carbonate budget for reefs starting out with higher coral cover and keep reefs with lower coral cover to begin with near equilibrium, at least in the short term. In the long term however, aggressive mitigations will need to take effect on a global scale for there to be hope of maintaining a carbonate budget near equilibrium.

This study is a report based on past studies of Caribbean reef systems and projections for the future under different scenarios using climate models. Kennedy *et al.* choose the Caribbean as their model system for multiple reasons—there has been ongoing research in the region on carbonate budgets throughout the time period that they are examining, carbonate budgets and ecological dynamics are simpler to model because of low diversity relative to other reef systems, and there have been severe disturbances of many sorts over the past decades that have been studied and can be used to predict and understand future trajectories of ecosystem functioning. Simulation models of climate change, ecosystem dynamics, and carbonate processes were run based on parameters drawn from published literature as well as some unpublished data on Caribbean reefs and climate data from IPCC AR5 earth system models. Interventions to reduce local stressors were separated from global efforts to mitigate greenhouse gas emissions to assess the action needed to sustain net carbonate production. The authors run models on reefs with a relatively healthy coral cover (~20%) and more degraded coral cover (~10%), and look at the effect of local protection of grazing parrotfishes. This analysis is considered under "business as usual" CO_2 conditions, as well as under a scenario where there is a move toward a lower carbon economy.

Major historical disturbances are outlined over the past several decades, beginning in the 1960s and 1970s, when the reefs had already experienced fisheries exploitation. This led to a boom in urchin populations because their natural predators were removed from the system, and this caused a net loss of reef structure due to an interaction between the urchins and crustose coralline algae. In the 1980s there was a region-wide die-off of branching coral because of disease, causing the most negative carbonate budget documented in the past decades. This event was followed by regional mass mortality in urchins, and the loss of these herbivores caused an abundance of macroalgae. In the 1990s and 2000s, bleaching events contributed to the overall lowered rates of carbonate production since the 1960s.

The authors found that under "business as usual" conditions, there was strong net erosion in the more degraded reefs regardless of local conservation efforts, and for reefs with relatively healthy coral cover, protection of parrotfishes caused a delay in the onset of erosion for approximately a decade. However, under the low emissions scenario, local conservation efforts were able to keep the carbonate budget at equilibrium for degraded reefs. When local parrotfish conservation measures were added to the low emissions scenario, there was a substantial increase in the carbonate budget. Thus they were able to conclude that it is necessary to begin with a relatively healthy reef for both local and global interventions to have a significant effect when the simulations were ended in 2080.

Grazers such as parrotfishes, and to a certain extent, urchins, are important to the ecosystem because they keep the macroalgae from dominating and provide space for coral recruits to settle. This implies that will be necessary to prevent overfishing of these grazers because coral cover declines in the functional absence of parrotfish, lowering the overall carbonate budget. Kennedy *et al.* suggest that carbonate budgets can be used to set target levels for coral, water quality, and herbivory to be better able to deliver biodiversity goals for reef management. This study suggests that local management and protection do have some preventative effect on the degradation of reefs, but that if we wish to see a lasting positive change in the carbonate budget or even carbonate budget equilibrium, dramatic global CO_2 emissions reduction measures will be necessary.

Conclusions

If current human practices are continued without change, the future does not look bright for coral reefs. Acidity does not appear to have as dramatic a direct effect on early life stages of corals or on the ability of corals to calcify as previously feared, however lowered pH can act as a resource for competitors that indirectly have a detrimental impact on the corals. Temperature appears to be a slightly more severe threat to the reefs, as corals have a harder time coping with warmer waters. Ocean acidification and warming are both clearly due to increasing levels of CO_2 in the atmosphere, and other local human activity increases the degradation of the reefs. Policy measures will need to be put in place on local scales if there is hope of slowing down the deterioration of coral reefs, and longer-term, larger-scale political action will be required to reduce CO_2 emissions if we wish to prevent the loss of one of the world's most beautiful, fragile, and dynamic ecosystems.

References Cited

Chua, C. M., Leggat, W., Moya, A., Baird, A. H. Temperature affects the early life history stages of corals more than near future ocean acidification. Marine Ecology Progress Series 475, 85–92.

Comeau, S., Carpenter, R., Edmunds, P., 2013. Coral reef calcifiers buffer their response to ocean acidification using both bicarbonate and carbonate. Proceedings of the Royal Society B: Biological Sciences 280, 20122374

Connell, S. D., Kroeker, K. J., Fabricius, K. E., Kline, D. I., Russell, B. D., 2013. The other ocean acidification problem: CO_2 as a resource among competitors for ecosystem dominance. Philosophical Transactions of the Royal Society B: Biological Sciences 368.1627

Gilmour, J.P., Smith, L.D., Heyward, A.J., Baird, A.H., Pratchett, M.S., 2013. Recovery of an isolated coral reef system following severe disturbance. Science 340, 69–71.

Howells, E. J., Berkelmans, R., van Oppen, M. J., Willis, B. L., Bay, L. K. Historical thermal regimes define limits to coral acclimatization. Ecology 94.5, 1078–1088.

Kennedy, E. V., Perry, C. T., Halloran, P. R., Iglesias-Prieto, R., Schonberg, C. H.L., Wisshak, M., Form, A. U. Carricart-Gavinet, J. P., Fine, M., Eakin, C. M., Mumby, P. J. Avoiding Coral Reef Functional Collapse Requires Local and Global Action. Current Biology 23, 912–918.

Munday, P.L, Dixson, D.L., Donelson, J.M., Jones, G.P. Pratchett, M.S., Devitsina, G.V., Doving, K.B, 2009. Ocean acidification impairs olfactory discrimination and honing ability of a marine fish. Proceedings of the National Academy of Science 106, 1848–1852.

Reyes-Nivia, C., Diaz-Pulido, G., Kline, D., Guldberg, O. H., Dove, S., 2013. Ocean acidification and warming scenarios increase microbioerosion of coral skeletons. Global Change Biology 19, 1919–1921

Venn, A., Tambutte, E., Holcomb, M., Laurent, J., Allemand, D., and Tambutte, S., Impact of seawater acidification on pH at the tissue–skeleton interface and calcification in reef corals. Proceedings of the National Academy of Sciences 110.5, 1634–1639.

9. Marine Pollution

Chloe Mayne

The world's oceans cover more than 2/3 of the earth's surface. These oceans contain a large diversity of life. Over the past 50 years, marine pollution has drastically increased and is adversely affecting the health of the ocean ecosystem and its organisms. Eighty percent of this pollution comes from the land based anthropogenic sources (UNESCO) and consists of plastic debris, toxic chemicals, and agricultural runoff. In addition, marine pollution comes in the form of noise pollution from sonar, boats, and ships. All of these sources cause a great amount of harm to the environment.

Every year around 220 million tons of plastic are produced (UNESCO). A large amount of this plastic ends up in the ocean and causes real problems. Large plastic debris affect animals through ingestion, suffocation, and entanglement, killing thousands of seabirds and marine mammals each year. The microplastics accumulate in organisms, causing issues with the endocrine and immune system. They eventually become part of the ocean food chain. Many of the plastics break down into small micro plastics, which leach toxic chemicals into the environment and do not biodegrade. The plastics also are pulled into gyres as a result of ocean currents, creating garbage patches such as the one in the Pacific Ocean.

Toxic chemicals not only are leached into the environment from plastics, but also come from agricultural runoff, residential runoff, oil drilling, treatment plants and many other sources. These toxins do not break down easily in the ocean environment. These include PCBs, DDT and pesticides as well as heavy metals such as mercury, lead and nickel. These toxins bioaccumulate in marine organisms and can cause real problems.

Nutrient-rich agricultural runoff is another type of marine pollution that harms many aquatic ecosystems. It causes eutrophication, which results in hypoxic water from an increase in algal blooms. An extreme case of eutrophication can cause a dead zone, wiping out all marine life in the area due to depleted oxygen.

Underwater noise is often not thought of as pollution, but may actually cause a great deal of harm to the marine mammals in an ecosystem. The noise may come from boat movement, military sonar, or oil drilling. Many marine species use sound to communicate underwater. The louder the ambient noise in the ocean, the harder for these animals to communicate in their environments. In many cases this may cause animals to abandon their habitats.

The Convention on Prevention of Marine Pollution by Dumping Wastes and Other Matter of 1972, known as the London Protocol, is the worldwide basis for preventing and controlling dumping in the marine environment. The United States followed suit with the Marine Protection, Research, and Sanctuaries Act, known as the Ocean Dumping Act. This act prohibits the dumping of material that degrades or causes harm to human health and the marine ecosystem. Although this act does quite a bit to prevent dumping, it is unable to prevent runoff or wind-blown debris from entering into the oceans.

It is clear that marine pollution has a negative effect on the ocean environment and its inhabitants. Scientific studies and experiments will help us to better understand the impact of the pollution and figure out what can be done to mitigate the current problems.

Marine Debris in Loggerhead Turtles: Bioindicator of the World's Oceans

Plastic marine debris has been accumulating at a rapid rate around the world. This debris, which comes from garbage, fishing, and ships, negatively affects marine environments and wildlife. Marine organisms, especially turtles, are harmed when they accidentally ingest this litter. Loggerhead turtles, *Caretta caretta*, feed generally on invertebrates on the bottom of the ocean, but have a high range of diversity in their prey. This makes it easy for the loggerhead to mistake plastic for food. These plastics carry many chemicals and pollutants, which result in higher mortality and possible negative effects on the entire food chain. The Mediterranean loggerhead population is isolated and has distinctive characteristics such as the small size of the adult turtles. This is important in relation to maturation and the time to regain population numbers because the Mediterranean basin has a large number of turtles caught each year as by catch. Campani *et al.* (2013) looked at the presence and abundance of marine debris in the gastrointestinal tract of the loggerhead sea turtles. They then worked to describe and quantify the types of plastic ingested by the turtles. Most of the turtles sampled had ingested marine debris, the most common type being user plastic, which is plastic used in everyday life .

Campani *et al.* used data from 31 stranded loggerhead sea turtles that were collected along the Tuscany coast. A necropsy (dissection) was performed on each turtle to determine cause of death. Each animal was weighed and measured and the most important organs were collected for chemical analysis. Following a prior bird marine litter determination protocol, Campani *et al.* determined the marine litter present in the loggerhead turtles gastrointestinal tract. The three main parts (esophagus, intestine and stomach) were divided up and separately analyzed by removing the contents, rinsing them and placing them onto dishes to be examined and sorted under a microscope. The frequency that each type of debris occurred was recorded as well as each specific weight of the sample. The results were then split into debris found in young, or juvenile, loggerhead sea turtles and adult loggerhead turtles.

This study found marine debris in 22 of the 31 sea turtles (71%) as there were a total of 438 pieces of marine debris. The majority of this was found in the turtles' intestine with less in the stomach and only one item, a hook, in an esophagus. The most debris found in the loggerheads was user plastic in 91.7% of them. One of the female loggerheads had the most marine debris of all the turtles with 90% being sheetlike user plastic. The small and young turtles had an average of 19 pieces of marine debris while the older and larger sea turtles had an average of 27 pieces. The analysis that was performed showed a relationship between number of plastics and shell length, shell length and plastic weight, and animal weight and plastic weight.

The presence of marine debris in the gastrointestinal tract of loggerhead sea turtles was found to be high in the turtles that stranded on the Tuscany coast. The greatest amount of marine debris found in the turtles was sheetlike plastic, which is commonly found in the ocean as a result of worldwide use and its long decomposition rate. The loggerhead sea turtle is an effective bioindicator of the marine litter in the environment and the health of the oceans.

Garbage Patch in the South Pacific Subtropical Gyre

The marine environment contains a large amount of anthropogenic plastic pollution. While the Northern Hemisphere subtropical gyres (NHSG) have been found to contain plastics, but so far there have been no data to suggest the existence of plastic pollution in the Southern Hemisphere subtropical gyres (SHSG).

Recently, a large amount of plastic pollution has been found in the Southern Pacific Ocean and along the coastal shores. It has began to negatively impact fishing, tourism and navigation. In addition to the South Pacific, large amounts of plastic have also been found in the Southern Ocean and near Antarctica. In order to look at the presence of microplastics in the South Pacific subtropical gyre, Eriksen *et al.* (2013) surveyed a transect that crossed directly through the gyre and took 48 samples. This transect was based on an accumulation zone created by currents and wind. The study found a greater amount of surface plastic pollution near the center of the transect than on the edges. These data prove the existence of a garbage patch in the South Pacific subtropical gyre.

Eriksen *et al.* organized an expedition to the Southern Pacific Ocean to collect data and explore the accumulation of plastic pollution. Along a 2424 nautical mile transect from Robinson Crusoe Island to Pitcairn Island, 48 samples of nueston, the organisms on top of the water, and associated debris were collected. Starting at 33°05'S, 81°08'W, samples were collected every 50 miles. After Easter Island, samples were collected every 60 miles until Pitcarin Island. A manta trawl was used to collect the sea surface samples. The net was towed on the side of the boat for a maximum of 60 minutes. The samples were stored in a 5% formalin solution. The sea state at the time of collected was calculated by looking at wave height and using the Beaufort Scale.

When placed in salt water, the plastic material floated to the top and under a microscope was removed. It was sorted into six size categories, which were then analyzed as to type of plastic: fragment, polystyrene fragment, pellet, polypropylene/monofilament and film. Of the 48 samples, 46 were found to contain plastics (96%). There were on average 26,898 pieces of microplastic per kilometer, weighing 70.96 g. A large majority of the count and weight was collected in the center third of the transect, which is where the plastic accumulates in the instance of a garbage patch. All six size classes were present in the samples and five of the plastic types were found. Fifty-five percent of the plastic particles were found in the size category of 1.00–4.75 mm. The most common type of plastic was fragments followed by plastic lines and films. The least common was plastic pellets although their total weight was still quite high. The sea state at each sample seemed to show an inverse relationship with the plastic count/weight; when the sea state was high, there was a low abundance of plastics and vice versa, probably because they were stirred into layers below where the manta trawl sampled.

This experiment indicates that the South Pacific gyre has a greater abundance of particles than the North Pacific gyre but says nothing about the abundance of plastics between gyres. This is due to the larger size of the SPSG, which skews the model. In reality, there are more plastics in the northern gyres as a result of greater amounts of production and consumption. The SPSG results are smaller than those of the NPSG, but are found to be within a similar magnitude. These results may vary depending on if there is an abundance of particles on the sea surface or below the surface. It is possible that microplastics wash up on shore, are ingested by marine animals or are driven deeper by waves and turbulence.

In these gyres, plastics form an accumulation zone as a result of surface currents that push the particles together. The plastic comes from rivers, beaches, maritime activities and illegal dumping. Overtime, it breaks down into smaller plastic particles as a result of wind and waves. Many of these plastics are petroleum-based and are considered a pollutant in the marine environment. Marine life is also affected by plastic entanglement and ingestion.

Plastic particles in the SPSG probably come from countries around the South Pacific Ocean, especially South America. Much of this plastic washes up on shore, but a large amount is also found to be in the South Pacific subtropical gyre. It is quite probable that some of the plastics are carried from as far as the South Atlantic and Indian Oceans. Cross equatorial transfer of plastics is also a consideration as some of the plastic may be from the North Pacific gyre. The results of this study prove that there is a garbage patch in the South Pacific filled with plastic pollutants. This experiment helps to predict the distribution of plastic particles around the globe and how they accumulate. With this information, it becomes possible to mitigate plastic pollution in the future.

Plastic Debris in Predatory Pelagic Fishes of the North Pacific

The North Pacific subtropical gyre contains large patches of marine debris and plastic. Recently, there have been reports of marine debris ingestion by sea birds, marine mammals, and fishes. Plastic debris is harmful to marine life, resulting in entanglement and decreased mobility, decreased nutrition or suffocation. Plastic also allows harmful organic contaminants to enter the marine environment, but there have been few experiments conducted on plastic ingestion in large marine fishes. Choy and Drazen (2013) studied 7 species of large pelagic fish for evidence of anthropogenic debris ingestion. Nineteen percent of the specimen had marine debris, primarily plastic or fishing line. A large majority of these species are thought to be mesopelagic fish that don't come close to surface waters where marine debris is usually found. Plastic in pelagic fish shows the possibility of plastic pollution making it's way through the food webs. These results are key in understanding the widespread nature of debris and plastic pollution in the ocean.

Drazen and Choy (2013) analyzed the stomach contents of 10 species of large pelagic fish collected from the central North Pacific Subtropical Gyre. The National Oceanic and Atmospheric Administration's (NOAA) Hawaii Observer Program used longline fishing to collect the specimen from 2007 to 2012. For each fish, the observers recorded the species, length, sex, and date of capture. The specimens came from shallow and deep depths on the longlines. The fish were randomly sampled and stomach samples were examined.

In the North Pacific, there are 2 species of opah, now known as *Lampris* sp. (big-eye) and *Lampris* sp. (small-eye). The *Lampris* species are normally not removed during fishing, so the stomachs for these species came from seafood wholesalers with capture date and location unknown. The remaining fish species were provided whole. The stomachs were opened and the contents were sorted into taxonomic levels. Items inside the stomach, both food and debris, were weighed and measured. The marine debris was assigned to one of five categories: white &

clear plastic, colored plastic, monofilament line, uncategorized line and rope, or uncategorized.

Since many of the species in the experiment live and feed deep in the water column, it was important to figure out where the pieces of debris were ingested. The buoyancy of the pieces was examined. Four seawater densities, based on temperature and salinity, were used to represent different depths in the gyre. Each seawater type was put into a glass beaker and every piece of debris was floated in each type of seawater for 60 seconds. Temperature and Salinity were measured throughout each trial.

The study found debris in 19% of the 595 individuals examined. Ten species of fish were studied, but only 7 species contained anthropogenic debris. The majority of the 262 total pieces of debris found were plastic, fishing line and rope. The two *Lampris* species had the highest percent of ingestion with the large-eye at 43% and the small-eye at 58%; the *Alepisaurus ferox* followed with 30%. The *Lampris* sp (small-eye) had the greatest number of pieces that were also the largest and widest. The most common debris color was clear/white or translucent and the second most abundant was black.

This study was one of the very first to focus on ingestion in large marine fishes as opposed to previous studies of micronekton fishes. With an extremely high sample size, there was a very high rate of debris ingestion for marine fishes. The three species with the greatest amounts of ingestion are considered mesopelagic fish, yet their stomach's contained plastic debris that are thought to come from the ocean surface. Either these fish are feeding on the surface at night without our knowledge or their prey is consuming the debris first. The latter is unlikely because the debris found is much larger than the pieces their prey would eat. Another explanation is the debris was consumed at depth rather than on the surface. The plastics may have become less buoyant and sunk by being waterlogged. The pieces may be drifting or slowly sinking in the water column. A final possibility is that plastic particles are pulled down below the surface as a result of downwelling, wind, or currents.

Why are these debris consumed by fish in the first place? The species with the lowest abundance of plastic and debris feed on the microscopic organisms on the ocean surface. In contrast, the species with the greatest abundance feed on gelatinous looking prey like amphipods, micronekton and squid. Most of these organisms are clear or white, which is similar to the large amount of clear/white pieces of debris found in this study. It is possible the fish mistake the plastics for their prey and ingest them by mistake.

The results of this study show the existence of plastics in the pelagic fish that humans consume most often and the existence of plastic pollution in the deep ocean food webs. The effect of plastic ingestion is unknown, but there are large amounts of PCB's, pesticides and metals that are absorbed in the plastic. For humans, this is a concern because large pelagic fish are part of our global food supply and may pose a toxic risk in the future.

Plastic Pollution and Associated Microorganisms in the North Pacific Gyre

Anthropogenic plastic pollution in the ocean has become extremely harmful to marine organisms and their environment. These problems include ingestion, entanglement, leaching of chemicals and adsorption of organic pollutants. The most common marine debris are small plastic fragments, which often are larger items that have been degraded. Microorganisms likely interfere with the degradation process as a result of biofilm formation on plastic surfaces. They may block plastic from UV radiation and photo-catalysis, which would increase plastic longevity. Inversely, microorganisms may accelerate degradation. Fouling microorganisms are extremely important to understanding the problems with plastic pollution, yet they have not been adequately studied. In this experiment, Carson *et al.* (2013) examine the abundance and diversity of microorganisms on plastics in the North Pacific Gyre.

Three important questions are addressed: what is the density of microorganisms on the surface of the ocean, how does the density vary across the Eastern North Pacific Gyre and how does the density vary in relation to plastic fragment properties of size, type, and roughness. The results found bacillus bacteria and pennate diatoms to be most abundant. The bacteria had an unevenly distributed abundance while diatoms had high densities on rough surfaces and where plastic concentrations were highest.

In July 2011, Carson *et al.* collected neustonic plastic in the Eastern North Pacific Gyre. The voyage on the R/VSea Dragon began in Honolulu, Hawaii and collected 17 trawl samples across the gyre in a northeastern direction. All of the trawl samples contained plastics. A manta trawl was used to collect the samples by towing the trawl for one hour alongside the ship. The exact trawl length was calculated using a flowmeter. To maximize the debris present on the surface and to avoid waves, the trawls were only utilized during sea states of 4 or less. The contents from the trawl were rinsed into a sieve and preserved with 5% formalin in a jar. Before and after each trawl, the sea surface temperature and salinity were documented.

Before examination, the samples were washed in water. From each trawl sample, eight fragments (1–10mm in size) were selected for analysis. Two of the trawls did not contain the correct size fragments, so only 100 pieces were used in the experiment. Each individual fragment was rinsed with ethanol, covered with a thin layer of gold, and examined under the Scanning Electron Microscope (SEM) for presence of microorganisms. A portion of each fragment was examined from top-to-bottom at 6000x magnification and the organisms were counted. Three random transects of the piece of plastic were analyzed and the data were recorded.

The fragments could not be aged and so each piece was identified as having either a "smooth" or "rough" surface. Species of microorganisms could not be identified so they were classified based on appearance. Diatoms were divided into three groups (pennate diatoms with a raphe, pennate diatoms without a raphe, and centric diatoms) while bacteria were divided into coccoid and bacillus. Dinoflagellates, coccolothopores and radiolarians were also recorded. The rest of the trawl samples were sorted and analyzed under dissecting microscopes. The plastic was dried and sorted by type before it was counted and weighed. Analyses of variance and linear regressions were performed to explain the abundance and diversity of microorganisms.

This experiment found a large number of plastic debris in the Eastern North Pacific Gyre. The trawls contained whole objects, fragments, pellets, line, foam, and films. The average number of pieces was 219 per trawl. Of the 100 pieces sampled, 59 were polyethylene, 33 were polypropylene and 8 were polystyrene. Each piece had microorganisms attached, the most abundant being bacillus bacteria and pennate diatoms. Coccoid bacteria, centric diatoms, dinoflagellates, coocolithophores, and radiolarians were found at much lower densities.

Bacillus bacteria density significantly varied with each trawl site and was also related to the type of plastic. Polystyrene fragments had the greatest abundance of the bacillus bacteria. A negative relationship was found between the bacillus bacteria and sea surface temperature, salinity, and longitude. High densities of bacteria were irregularly distributed within the gyre, while diatoms had the highest densities in the center of the transect. Trawl site and diatom density were significantly related. Density was also high on rough fragments and sites where there was more plastic. Diatoms had a significant negative correlation with temperature and salinity. As the transect moved northeast, density of pennate diatoms with a raphe decreased while those without a raphe increased. Centric diatoms were extremely low in density in this experiment.

The average number of fragments found in this experiment was larger than in the South Pacific and North Atlantic Gyres. While the concentration is highly variable, there was an increase of plastic in the center of the transect. The distribution of plastic-associated organisms was quite irregular. Some variation was explained by the number of microorganisms found on the specific type of plastic. Diatom abundance peaked in the center of the gyre along with the plastic concentration. Higher amounts of plastic may make it easier for these microorganisms to colonize the new surfaces. Plankton abundance did not have a similar distribution. Since there was no standardization during the trawl collections, it is possible that vertical migration changed the plankton abundance based on morning or afternoon trawls.

Surface roughness was related to a high diatom density but not high bacteria density. Rough surfaces are a sign of plastic degradation, which may speed up as a result of colonization. This could result in the sinking or ingestion of an item, which becomes part of the food chain. This process may be accelerated in the gyre center, as there is a greater abundance of plastics and microorganisms.

With global plastic production increasing rapidly, it is expected that there will be a large amount of plastic debris on the ocean surface. Future research is important to understand what can be done once the input of plastic debris slows or stops. Degradation of plastics in the ocean, though not understood, is most likely a result of microbial biofilms. The future of plastic pollution is based on understanding and identifying the microorganisms on the ocean surface and the effects they have on the debris.

Plastic Ingestion in Mediterranean Sperm Whale

Marine debris has become an extensive problem in the oceans today. Plastics and other debris affect over 250 species of marine organisms by entanglement and ingestion. Current studies have shown large marine mammals to be affected by entanglement, but few studies exist on the affects of ingestion of plastic debris. There have been three recorded standings of sperm whales with large amounts of plastic and marine debris in their stomachs. In the Mediterranean Sea, the population of sperm whales is considered to be a separate species and has had a decline in population over the past 20 years. This decline has been attributed mostly to boat strikes and entanglement, with no knowledge of the affects of plastic ingestion. Stephanis *et al.* (2013) examined a stranded sperm whale in the Mediterranean Sea that had ingested a large amount of marine debris. They discuss the results in terms of the spatial distribution of sperm whales and the anthropogenic activities in the area. In the Mediterranean Sea, the species is found near Almería and the Strait of Gibraltar. This study found that the whales feed in an area with lots of debris from the greenhouse industry. In the sperm whale examined, the cause of death was determined to be gastric rupture from the build up of debris.

On March 28th, 2012, a male sperm whale stranded near Castell De Ferro, Granada. The whale was weighed and measured. The abdominal cavity was also opened and the contents of the stomach were recovered. They were taken to the Estación Biológica de Doñana-CSIC in Spain where they were labeled, weighed, and measured. For pieces of anthropogenic material greater than 4 cm^2, type of material, diameter, color, and origin were determined. This helped to categorize the items into either greenhouse debris or general debris. Items less than 4 cm^2 were put together and considered small plastics. The sperm whale's fluke was compared with a catalogue of sperm whale flukes assembled by Carpinelli *et al.*

Data about the spatial distribution of sperm whales were collected through surveys from research projects of the NGO Alnitak from 1992 to 2009. Over the survey period, 74,187 km were sailed and 34 groups of sperm whales were recorded with 55 individual whales. The Alboran Sea was the study area and it was characterized by spatial and environmental variables such as depth, slope, sea surface temperature, and primary production.

This study found the stranded whale to be 10.0 m long and weigh 4500 kg. The sperm whale was extremely emaciated but there was no evidence of other injuries. The abdominal cavity of the animal was opened for inspection. The stomach contained squid beaks and a large mass of compacted plastics that ruptured through the first stomach compartment. There were squid beaks on the outside of the small intestine but nothing was found to be inside the intestines. Gastric rupture was considered the cause of death. The stomach contents that were recovered were mostly greenhouse and general debris. The most common debris found was the cover material of greenhouses. There were 64 pieces of greenhouse debris and 21 pieces of general debris. The sperm whale's fluke had not been photographically matched in the Mediterranean Sea.

While marine debris has been encountered in sperm whale stomachs throughout the 20th century, none of the debris has been attributed to greenhouse activities. In the Mediterranean Basin, greenhouse agriculture has expanded rapidly as people are able to grow year-round. These greenhouses require many plastic materials in their construction. The most common debris found in the sperm whale was the plastic cover of the greenhouses. There were also two flowerpots and seven bits of burlap plastic bags found in the whale's abdominal cavity.

There was a bimodal sperm whale distribution based on longitude, with a high density toward the Straight of Gibraltar and another small peak around Almería. The model for group sizes showed that they increased in density towards shallower waters and it can be seen that the animals are feeding in waters near an area with a large greenhouse industry. There are no data of floating debris in the Mediterranean Sea, making it difficult to know where the debris resides in the water column and how much plastic is in the ocean, but it is likely that it came from greenhouses on shore.

This study only examined one stranded sperm whale since it is quite difficult to recover a sperm whale carcass unless it is washed to shore and hasn't fully decayed. While there may be few other studies, it is obvious that addressing the issue of plastics from greenhouse agriculture in the sea is important.

Record of Contaminant Exposure Found in Blue Whale Earplug

The blue whale is the largest animal on earth and an Endangered Species. The blue whale population has not only been affected by whaling activities but also by anthropogenic activities such as fishing entanglement, boat strikes, climate change and chemical emissions. Many of the resulting contaminants of the latter are found to reside in blubber and other lipid rich layers. Blue whales, and other large baleen whales, accumulate layers of earwax, known as cerumen, throughout their lifetimes. This wax builds up over time and produces dark and light colored layers, or laminae, which represent periods of feeding or migration. While early earplug analysis was used to age whales, it was recently discovered that it is possible to reconstruct lifetime profiles of hormones, mercury, and other pollutants. in whales. Similar to blubber, these contaminants build up in the whale's earplug in large enough concentrations to quantify. Tumble *et al.* (2013) looked at the earplug of a male blue whale to examine lifetime contaminant and hormonal profiles. They found 24 different laminae and estimated the whale's age to be 12 years ± 6 months. Looking at the hormone levels, they were able to age the whale's sexual maturity. When the contaminants were examined, it was found that much of the contaminants came from mother during the first 12 months. The mercury profiles showed peaks as a result of environmental and anthropogenic mercury increase.

Trumble *et al.* obtained an earplug from a 21.2 m male blue whale that was killed by a ship strike in 2007. The earplug was cut longitudinally with a saw to make the internal laminae more accessible. The earplug was found to have 24 distinct laminae. Each layer was examined under a microscope, removed from the section and stored in nitrogen at –30°C . Next, the compounds were analyzed. Hormone analysis was completed using the Enzo Life enzyme kit for each respective hormone. Mercury was determined using the mercury analysis system approved by the EPA. The organic contaminants (POPs) were determined using a pressurized liquid extraction method.

The findings of this experiment looked at the lifetime profiles of cortisol (stress related hormone), testosterone (development hormone), POPs (persistant organic pollutants), and mercury. During the lifespan of the blue whale, cortisol concentrations doubled. Cortisol concentrations are a biomarker of stress in mammals and are in response to environmental, physical, and chemical factors. The large peak in cortisol concentration may have been related to breeding competition or social bonds, while the general increase during the animal's life was due to migration, food, environmental conditions or contaminant exposure.

The results for testosterone showed a large peak at around 10 years, indicating sexual maturity. The testosterone levels ranged from a low of 230 to a high 93,000 pg/g. Of the 42 POPs analyzed, only 16 were measured in the cerumen. The POP concentrations peaked at 0–6 months and then decreased. This result showed that there was a large amount of maternal transfer during gestation & lactation. Around 20% of the whales lifetime burden can be attributed to maternal transfer. The lifetime burden is considered the uptake, metabolism & excretion of POPs. During feeding and fasting, these POPs are recirculated throughout the body but their effects are unknown.

The mercury profile showed two distinct peaks found at 60–72 months and 120–126 months. These peaks may be due to increases of environmental or anthropogenic mercury during the blue whale migration along the coast of California. Mercury in the environment can be harmful as it bioaccumulates and causes impaired neurological development.

This experiment was successfully able to create a lifetime profile of an individual blue whale by measuring the trace amounts of contaminants and hormones from the layers that build up in the cerumen of baleen whales. Using the earplug is much easier than analyzing blood, feces and blubber. In addition, earplugs provide a more comprehensive examination of the contaminants, hormones and persistent compounds. Earplugs are extremely advantageous because it is also now possible to analyze samples that were taken as early as the 1950s. Finally, examination of whale earplugs provides the opportunity to understand the anthropogenic effects on not only an individual whale, but the whole marine community.

Mid-Frequency Military Sonar Affect on Blue Whales

Anthropogenic underwater noise, has recently been found to have negative effects on marine life. While there have been mass strandings of deep-diving toothed whales due to mid-frequency military sonar, the effect of military sonar on baleen whales, like the blue whale, is currently unknown, but there have been a few cases of baleen whale strandings that showed a common pattern of gas-bubble lesions and fat emboli from changed diving behavior. Sound exposure may cause behavioral changes leading to disorientation, injury, stranding and eventually mortality. Goldbogen *et al.* (2013) conducted controlled exposure experiments on 17-tagged blue whales in Southern California to look at their response to mid-frequency sonar. The results demonstrated that blue whale behavior is significantly affected by mid-frequency sound. The behavioral changes seen consisted of the cessation of deep feeding, increased swimming, and travel away from the sound. The disruption of feeding and movement away from dense patches of krill may have negative impacts on baleen whale foraging, fitness, and population health.

Goldbogen *et al.* performed controlled exposure experiments (CEE's) to measure the effect of anthropogenic sound on blue whale behaviors. They tagged 17 blue whales off of the Southern California coast during summer and autumn 2010. The digital tags contained sensors for body orientation, swimming activity, depth, speed, and received sound levels. The variables were then divided into dive behavior, body orientation and horizontal movement. Sixteen whales received DTAGs while only 1 whale had a Bioacoustic Probe. For each tagged whale, the research consisted of three 30-minute periods: a pre-exposure period to find baseline data, an exposure period, and a post-exposure monitoring period. The sound source from the research vessel simulated actual military sonar (MFA sonar) signals and pseudo-random noise (PRN) intended to imitate the US Navy's sonar at lower levels. The signals were projected from a 15-element vertical line array of active transducers.

The whales were exposed to either MFA or PRN, which was transmitted at a starting level of 160 decibels (dB) at 1 meter. Every 25 seconds for 30 minutes the sound source generated a signal that was ramped up by a 3dB increment. The MFA was 1.6 seconds long and had a maximum sound level of 210 dB while the PRN was 1.4 seconds long and had a maximum of 206 dB.

Behavioral and environmental variables from each whale dive were recorded and analyzed. These consisted of a wide range of measurements including maximum depth, dive duration, descent time, bottom time, ascent time, surface time, and average speeds during descent/ascent. The behavioral state from the tag data was divided into three categories: deep feeding (below 50 m), surface feeding and non-feeding. Statistical analyses were conducted on the 54 variables which allowed the researchers to assess the significance of the behavioral response and if there was a difference between MFA and PRN responses.

The result of this experiment showed that whales' diving behavior was significantly affected by exposure to mid-frequency sound. Surface feeding whales were not noticeably affected while deep feeding and non-feeding whales responded with termination of deep diving and longer mid-water dives. Sound type also resulted in varied responses, but many of the behavioral changes returned to the baseline behavior once the exposure time was over. The behavioral responses to acoustic signals were extremely complex and depended on different variables. Some of the responses may be seen as avoidance responses since behavior returned to pre-exposure conditions after playback ended. Surface feeding blue whales do not deal with diving costs and thus their responsiveness to sound is decreased. Whales on the surface also had a lower maximum dive level achieved because of possible reduced sonar levels in shallower depths. The results show behavioral variation is due to interactions of behavioral state, environmental factors and individual differences in whales.

An important piece of data showed that one of the blue whales stopped foraging and swam away from the source when the sound exposure began. Blue whales need to forage on krill throughout the day in order to support themselves. With the ceased feeding, the whales' foraging efficiency is lowered. This specific whale went from eating 19 kg of krill per minute to eating no krill for 62 minutes, a loss of 1 metric ton of krill during the sound exposure. This is extremely costly for the whale, as its body needs this great amount of food.

Repeated sonar exposures could have a negative effect on feeding performance, body condition, fitness and even population health. The simulated sonar signals used were not as intense or as long as the real ones in military systems. The actual effects of sonar may have longer responses from whales and may be dangerous to the endangered blue whale populations in North America.

The result of this experiment shows that blue whales are responsive to mid-frequency sonar but their responses are complex and are based on behavioral and sound exposure factors. In most cases, they involved avoidance of the sound exposure until it ended and behavior returned to normal. This study makes it clear that it is important for decisions on military sonar to be considered in relation to their negative effects on blue whales.

Chloe Mayne

Displacement of Bottlenose Dolphins from Anthropogenic Boating Noise

The Cres-Losijn archipelago and the surrounding waters receive a large amount of boat traffic each year due to leisure boating during the busy tourist season. The boat presence may be affecting the environment of the common bottlenose dolphins in the area. Between 1995 and 2003, the local population showed a decrease of 30%. The boats' impact is due to either physical harassment or acoustic disturbance from the loud engines. This anthropogenic noise sends a low frequency input of energy into the water, which increases the sea ambient noise (SAN). Bottlenose dolphins rely on sea ambient noise to communicate, navigate, and forage in their environment. Boat noise can affect the dolphins by modifying their acoustic behavior, surface behavior and diving intervals. The study by Rako *et al.* (2013) examined the change in underwater noise levels from leisure boating and the effect on bottlenose dolphin distribution in the northern Adriatic Sea. Sea ambient noise was collected at 10 acoustic stations over a three-year period. During this period, data were collected on the presence of leisure boats and the spatial distribution of bottlenose dolphins. The results of this study show a correlation between high SAN levels and boat presence. There was also seasonal displacement of dolphins during the noisy leisure boating seasons.

Rako *et al.* (2013) studied a 545 km^2 area in the Kvarneric region of the northeastern Adriatic Sea. This area has steep rocky shores and a seabed that averages 70 m deep. It is a well-known tourist destination in the northern Adriatic Sea, with the largest number of tourists visiting during the summer. The data for the experiment were collected from January 2007 to December 2009 from sea ambient noise (SAN) monitoring on 10 acoustic stations. The recordings were projected using a RESON TC4032 hydrophone and a Pioneer DC-88 DAT recorder. The hydrophone was lowered to 4 meters deep from an inflatable boat and each recording lasted 5 minutes. The acoustic stations were grouped into three categories of high, medium and low anthropogenic impact. These classifications were based on the stations proximity to urban centers, tourist destinations and boating routes.

During the three-year period, 418 samples were recorded, with each station having a mean of 42 recordings. The recordings were analyzed for their frequencies in terms of Sound Pressure Level (L_{eq}). The L_{eq} was also calculated for two frequency ranges. The first was between 63 Hz and 2 kHz, where there is a high contribution of anthropogenic noise, and the second was from 2–20 kHz, where natural sound sources are expected.

Since this study focused on the presence of leisure boating in the area, data were collected within 2 km of the acoustic station using marine binoculars. This distance was to ensure accurate estimation of boat type and to avoid replication in the boat count. The boats present were then analyzed based on season and spatial orientation in the high, medium and low impact areas.

Dolphin presence was collected for this experiment during favorable sea conditions and fair visibility. Garmin GPS was used to record the coordinates of each dolphin encounter. Group size and age class for each dolphin sighting was estimated based on body size, color and behavior. Groups that were encountered within 2km from an acoustic station were analyzed based on seasonal changes in number of encounters. Dolphin presence was analyzed by dividing the study area into 1000 x 1000 m cells and finding the average cell encounter rates (ERs).

The results of this study found that overall sea ambient noise in the area did not vary significantly during the study period. Across the acoustic stations and between the two seasons, noise changed significantly for the 63 Hz to 2 kHz range. The SAN was higher during the tourist season and changed differently at each acoustic station. An ANOVA showed that noise in the low frequency range didn't change between the high, medium and low impact areas but the affect of season in these areas was significant. During the tourist season, the high impact area had higher SAN levels than the low. The number of boats increased significantly in the study area with 302 boats recorded during the tourist season and 61 in the off-season. The highest number of boats during the tourist season was found to be in the high impact area, while the greatest number of boats during the non-tourist season was found to be in the medium impact zone. A positive correlation was found between SAN levels and leisure boating for both frequency ranges. There was also a significant seasonal correlation during the tourist season between noise and boat presence.

Over the study period, 170 bottlenose dolphin encounters were recorded. Seventy-five encounters were during the tourist season and 95 were from the non-tourist season. There were higher encounters during the non-tourist season in the center of the study area and during tourist season at the edges of the area. The tourist season had the highest number of encounters at stations 5, 7, and 8, which corresponds with the lowest SAN levels. In addition, only one dolphin encounter occurred in the high impact area during the tourist season, while many were there during the off-season.

Monitoring of boat traffic during SAN sampling shows the contribution of anthropogenic noise in Adriatic Sea. The large amount of boats increased the sea ambient sound, which peaked for low frequency ranges during the tourist season. Leisure boating may greatly affect the habitat of the local bottlenose dolphins as a result of anthropogenic disturbance. High numbers of dolphins were spotted during the non-tourist season as a result of decreasing human pressure. In addition, a greater number of dolphins were found in the areas that had the lowest sea ambient noise. This may be due to the dolphins' avoidance of noisy areas as they use sound for their communication and livelihood. Noise disturbance may result in increased stress and abandonment of habitats for the bottlenose dolphins. This area in the Cres-Losinj archipelago is an important nursing ground for the animals, but with increased boating activities they may be displaced and lose time for females to nurse their calves.

Noise pollution has a negative effect on dolphins, as sound is the main sense they use. In the long-term, boating noise may drive populations to completely abandon their preferred habitats. It is important to develop conservation measures related to boating limits to mitigate the impacts of leisure boating on the bottlenose dolphin population and habitat.

Conclusions

Marine pollution is an issue that needs to be addressed immediately as it causes harm to an extremely important and fragile environment. Much work will need to be done in order to prevent marine pollution from occurring and to clean up the debris and toxins that are in the oceans today.

References Cited

Campani T., Baini M., Giannetti M., Cancelli, F., *et al.*, 2013. Presence of plastic debris in loggerhead turtle stranded along the Tuscany coasts of the Pelagos Sanctuary for Mediterranean Marine Mammals (Italy). Marine Pollution Bulletin 74, 225–230.

Carson, H., Nerheim, M., Carroll, K., Eriksen, M., 2013. The plastic-associated microorganisms of the North Pacific Gyre. Marine Pollution Bulletin 75, 126–132.

Choy, C., Drazen, J., *2013*. Plastic for dinner? Observations of frequent debris ingestion by pelagic predatory fishes from the central North Pacific. Marine Ecology Progress Series 485, 155–163.

Eriksen, M., Maximenko, N., Thiel, M., Cummins, A., *et al.*, 2013. Plastic pollution in the South Pacific subtropical gyre. Marine Pollution Bulletin 68, 71–76.

Goldbogen, J., Southall, B., DeRuiter, S., Calambokidis, J., *et al.,* 2013. Blue whales respond to simulated mid-frequency military sonar. Proceedings of the Royal Society 208, 1–8.

Rako N., Fortuna CM., Holcer, D., Mackelworth, P., *et al.,* 2013. Leisure boating noise as a trigger for the displacement of the bottlenose dolphins of the Cres-Losinj archipelago (northern Adriatic Sea, Croatia). Marine Pollution Bulletin 68, 77–84.

Stephanis, R., Gimenez, J., Carpinelli, E., Gutierrez-Exposito, C., Cañadas, A., 2013. As main meal for sperm whales: Plastics debris. Marine Pollution Bulletin 69, 206–214

Trumble, S., Robinson, E., Berman-Kowalewski, M., Potter, C., Usenko, S., 2013. Blue whale earplug reveals lifetime contaminant exposure and hormone profiles. PNAS published ahead of print September 16, 2013,doi:10.1073/pnas.1311418110.

UNESCO. Facts and figures on marine pollution. http://www.unesco.org/new/en/natural-sciences/ioc-oceans/priority-areas/rio-20-ocean/blueprint-for-the-future-we-want/marine-pollution/facts-and-figures-on-marine-pollution/

Section III. Terrestrial Issues

10. Agriculture and Climate Change

Caroline Vurlumis

Currently the world's population is over 7 billion people and this number increases daily. The global population has only been able to reach this enormous number thanks to continuously developing agricultural methods dating back to the industrial revolution. As a result, the human race has become increasingly dependent on industrialized agriculture to feed an exponentially growing population. Such advanced and extensive agricultural methods (defined as the cultivation of crops as well as livestock) and the conversion of land however, contribute to harmful greenhouse gas emissions. Nitrous oxide (N_2O) and methane (CH_4) are released through various soil management practices, livestock, manure management, and rice cultivation and contribute to the overarching problem of climate change. Furthermore, conversion of forest to agriculture releases large amounts of carbon dioxide (CO_2) from deforestation. Additionally it also harms biodiversity. When land is converted to agriculture, watersheds and groundwater are susceptible to pollution from fertilizer runoff. This not only leads to impacting species but also affects human health.

Agriculture is extremely sensitive to climate change. As more greenhouse gases are released into the atmosphere, temperature, seasons, weather, and precipitation are being affected. Rising temperatures mean more insects, weeds, and crop diseases, and in some parts of the world will disable agricultural cultivation altogether, but some parts of the world nearer the poles such as northern Europe, Canada and Russia may experience some improvement to agriculture. These minimal gains however are not likely to outweigh the losses elsewhere. Climate change has already had a huge impact on yields in maize, wheat and other major crops. The decrease in agricultural productively will have a major impact on a demanding growing population.

In order to investigate the impacts of climate change on agriculture there are scientific studies being conducted all over the world. This chapter presents just a sample of current scientific research that was published in 2013. These studies derive from Brazil, USA, China, Poland and Nigeria—they cover a variety of topics including modeling future agricultural suitability, impacts of ecosystem conversion, thoughts on climate change, sustainable intensification, biofuel crops, nitrogen fertilizer, agricultural management, and hydraulic changes in watersheds. Although knowing and understanding the science behind impacted agriculture is important, it is just as vital to consider policy and farmer beliefs when addressing climate change.

Predicting Climate Change Effects on Agriculture

Anthropogenic climate change may lead to more extreme weather events and will impact human health and the economy. One of the most important effects of climate change however is its impact on agriculture and the human population. Using ecological niche modeling (ENM), a technique to relate species presence to environmental factors, Beck (2013) sought to model suitability of agriculture in different regions based on soil conditions and climate. He applied this general and simple model for agriculture across the Old World (defined as Asia, Africa and Europe) and the Australia/Pacific region to create model scenarios of agriculture for the year 2050 and determined which countries would win or lose from climate change in the agricultural sector; different regions vary considerably in agricultural suitability which cannot simply be determined by a country's wealth. Beck's model predicted that parts of Europe, Africa and southern and eastern Asia will have a negative impact while north-eastern Europe and the Tibetan plateau will benefit.

Beck started his modeling by first obtaining 290 records of agriculture presence worldwide. These data were assembled from 'traditional' plants (excluding cash crops, pastures, greenhouses and high-tech practices) and used to make a probability model of agriculture occurrence from current climate and soil conditions across the Old World. The sites chosen contained staple crops such as maize, rice, wheat and rye. Current data taken from www.worldclim.org and data from the Food and Agriculture Organization (bioclimatic and soil types) were applied in Beck's ENM. The maximum entropy method was applied to determine the best data set for accessing the probability of future agriculture change. In addition, the author used two global circulation scenarios: the A2a (economic interests followed, fast population growth) and B2a (social and environmental interests followed, slower population growth), which represent a negative and a positive scenario of climate change and economic growth patterns. From these models and data the author predicted agricultural suitability possibilities in 2050.

Beck's models predicting suitability of agriculture presented in the global circulation models A2a and B2a both foresaw a decrease. There was variation but overall the loss was high. There were barely any losses in rye growing regions, intermediate losses for wheat and maize, and larger losses in tropical plant areas. East Europe on average was predicted to benefit from climate change while West Africa was predicted to suffer. Other regions, with what is currently judged to be agriculturally poor land including Western Europe, will benefit from climate change. In general, results reflected similar projections from earlier studies showing slight decreases for agriculture worldwide and large variation in results.

Since ENM is a simple approach that uses less data on productivity and agricultural practices than previous models, the author acknowledges the potential errors that could have been involved, including input data error, technical issues, and misinterpretation. This model assumes local variation of agricultural practices in addition to local adaptation. Nevertheless, the data represented by this ENM model suggest that both rich and poor countries will suffer loses in agricultural suitability and this model should be used in accordance with other models to predict future trends.

Iowa: Farmer Thoughts on Climate Change

Agriculture is being threatened by climate change and yet feeds this threat by emitting greenhouse gases itself. In Iowa, the site of some of the most productive land in the world, a survey investigated farmer perceptions of climate change and their response to altering farming practices. The two major research questions Arbuckle *et al.* (2013) investigated were: 1. Do farmers support adaptation and mitigation actions and 2. Do beliefs and concerns about climate change influence those attitudes? The surveys supported the authors' hypothesis that farmer's level of concern about how climate change would impact their livelihoods was correlated with mitigation strategies. Farmers who did not believe climate change was human-caused or was not a problem were more likely not to support mitigation.

This study took place in Iowa, which is known as the "Corn Belt" for producing half of the corn and soybean supply in the US and is where extreme weather attributed to climate change has become increasingly frequent. These changes include warmer winters, longer growing seasons, and higher dew point temperatures. These shifts impact agricultural production and the natural resource base. Arbuckle *et al.* initiated their study with questions in mind of whether farmers support adaptation and mitigation actions in response to climate change and whether beliefs and concerns about a changing climate influences attitudes. The authors held several hypotheses labeled as H1, H2a, H2b, H3a, and H3b regarding the relationships between attitudes towards adaption to climate change and its impacts. They predicted that farmers who expressed concern and thought climate change was real would mitigate farming practices if it proved effective while those who did not believe change was occurring would not change their methods. The independent variables in this study were a concern scale and ingenuity. The dependent variables included the three actions of protection, drainage, and mitigation, in response to climate change. The three controlled variables were acres of row crops farmed, age, and education. The surveys asked on a scale of five classifications (strongly disagree, disagree, uncertain, agree, strongly agree) what farmers beliefs on climate change were (concern and ingenuity) and what their opinions were on protection, drainage, and mitigation. Arbuckle *et al.* used a multivariate statistical model to identify relationships between beliefs about the existence of climate change, impact concerns, and attitudes towards different responses.

The results showed that level of concern and age were significant indicators of whether farmers would take action to protect farmland from damage. Farmers who expressed more concern were more likely to take additional steps to alleviate adverse effects of increased precipitation—these results were consistent with hypothesis H1 (concern positively associated with support for protection, drainage, and mitigation). Hypotheses H2a and H3a, predicting ingenuity and belief about climate change would be positive predictors to support action, were not supported. Concern and ingenuity were found to be strongly associated with support for additional agricultural drainage consistent with H1 and H2a (ingenuity positively associated with protection and drainage). Belief in climate change however did not prove to be a positive predictor. Concern was found to be a positive predictor of government action while ingenuity was found to be a negative predicator of mitigation. Just as the authors expected, the relationship between belief in anthropogenic climate change and applying mitigation methods had a positive relationship. The farmers who believed climate change was caused naturally or were uncertain of its existence were less likely to support government mitigation actions.

This study was one of the first efforts to investigate the relationship between farmer belief on climate change, concern, and attitude towards adaptation and mitigation in agriculture. Arbuckle *et al.*'s overall hypothesis of the surveys was supported. Farmer beliefs of climate change and their concern for its impacts to them personally directly influenced their response to adaptation and mitigation. These surveys indicated how important perception is to climate risk on one's farming methods. It is not influential enough for one to believe in climate change but farmers must understand it is largely caused by man and will affect them directly. Farmer's views on climate change are critical to maintaining agricultural productivity and resilience, especially if policymakers want to create effective, agriculturally protective changes.

Bacterial Communities After Ecosystem Conversion

The Amazon rainforest is the most extensive tropical forest worldwide containing the largest amount of plant and animal diversity. For years there has been deforestation to create agriculture and cattle pasture which greatly harms biodiversity and causes homogenization in bacterial communities. Rodrigues *et al.* (2013) tested the impacts of soil microbial biodiversity when land is converted from forest to agriculture. Using transects from forest and pasture, the authors took soil samples of each to test local (alpha) diversity and differentiation (beta) diversity by concentrations of taxonomy and phylogeny. The results showed an increase in alpha diversity and decrease in beta diversity indicating a significant difference in bacterial communities when conversion occurs. There was significant loss in endemic species diversity and an increase of homogenization in the soil which poses a higher risk for net loss of biodiversity in the future. As a result of this study the authors argue that microbial biodiversity loss should be strongly considered when engaging in land conversion due to its important role in tropical ecosystems.

This study took place at Fazenda Nova Vida in Rondonia, Brazil. In a previous study bacterial diversity was not depleted during conversion of the Amazon but these findings were based on a limited sampling depth and local diversity. To prove that animal and plant diversity actually decrease after ecosystem conversion, Rodrigues *et al.* designed a study without these limitations. They took 10-km transects of both forest and pasture and from these used the spatially nested scheme to take 36 samples of soil from each. Bacterial communities were characterized by barcoded pyrosequencing and then put into taxonomic units. Community diversity and similarity were calculated for both taxonomic and phylogenetic traits. Based on the samples, the authors estimated the local (alpha) diversity, and differentiation (beta) diversity between the soil samples.

The results showed that the forest and pasture soil samples contained significantly different amounts of taxonomic and phylogenetic structures. Converting Amazon forest land to agricultural land resulted in a large decrease of the phyla *Acidobacteria*, *Nitrospirae* and *Gemmatimonadetes*. The phyla *Firmicutes*, *Actinobacteria* and *Chloroflexi*, on the other hand, all increased during conversion. When comparing environmental characteristics, forest soils and their microbacterial communities had lower pH, carbon, and more aluminum, while pasture communities increased in carbon, nitrogen, and magnesium concentration. Overall there were significantly more bacterial communities found in the pasture soils than in the forest soils.

After conversion, bacterial communities were significantly altered between forest and pasture. The decrease in differentiation (beta) diversity indicates increasing similarity in the community (biotic homogenization). The loss of endemic taxa and increase in similarity between phylogenetic bacteria suggests a higher risk of biodiversity decline during conversion. A higher homogenization increases risk since there is less resilience to disturbance which could lead to a net loss of diversity. As more soil is sampled, the forest surpasses pasture samples in taxonomic richness. To fully argue this, more sampling needs to be done to document diversity loss. In this study there was a minimal difference detected but it nevertheless represents a huge loss of genetic variation after ecosystem conversion. There was a significant reduction in forest endemic species while homogenization of taxa increased. A change of microbial diversity through conversation should be of concern since microbes play an important function in the ecosystem. Further studies need to be conducted to find out the long-term impacts of biodiversity loss.

Caroline Verlumis

Poland: Impacts on Farmland Birds After EU Accession

Identification and abandonment of farmland are two major changes that have occurred in the European Union (EU) agricultural management system. In 2004, Poland became a Member State of the EU. As 70% farmland, Poland supports more than 11% of farmland birds in the EU making it a significant contributor to the farmland bird index. In 2002 and 2003, before Poland became a member of the EU, there were 180 1-km squares of agricultural intensity surveyed of bird and farmland habitat. In 2009, after accession into the EU, 71 of these plots were resurveyed to assess the impact on farmland bird populations and to quantify intervening agricultural change. Sanderson *et al.* (2013) tested a series of explanatory variables to examine whether any changes in agricultural management were correlated with bird species abundance and richness. The authors discovered that ever since Poland's accession to the EU, new agricultural management methods such as those included in the Common Agricultural Policy (CAP) have had detrimental effects on bird populations. In this study the abundance of 9 out of 28 bird species decreased due to loss of low-intensity farmland.

Sanderson *et al.* first surveyed birds and farmland habitats in three regions in Poland in 2002 and 2003. After being a member of the EU for five years, the authors randomly selected 71 of the original 1-km squares and only squares that were surveyed in both years were considered in the analysis. Each square was surveyed twice for bird species with a testing time lasting between 2–3 hours. All birds that were seen or heard were marked on a 1:10,000 scale map and the route was marked so it could easily be replicated. The land cover types taken into consideration included arable, recently abandoned, set-aside, unused farmland, and grassland. These were all assessed to determine how the loss of low-intensity farmland and similar habitats affected bird species and populations.

Changes in habitat included low intensity farmland cover, total farmland cover and arable land cover, and were analyzed using linear mixed models. Bird population density in relation to habit change was also analyzed using a binomial distribution with logarithmic links. Explanatory variables included percentage change in total farmland cover, change in grassland and abandoned land, change in wood edge length, interaction between change in grassland and abandoned land and direction of change in fallow land. Furthermore, these explanatory models were compared with all bird species, farmland specialists, ground nesting farmland specialist and tree/edge nesting farmland specialists.

In the three Polish regions the overall cover in arable land increased in one region, the low cover intensity farmland decreased in two regions, the total farmland area cover decreased in two regions and the wood cover increased across all regions. Out of the 28 species that were examined, the most significant habitat change was change in low-intensity habitats; abandonment did not have as much of an impact. The species to experience a significant decline included meadow pipit, European greenfinch, Eurasian tree sparrow and the marsh warbler.

Since its accession to the EU in 2004 both intensification and abandonment have affected Polish farmland birds. This study was the first chance the relationship between bird abundance and habitat change could be assessed pre and post-accession into the EU. The results indicating a loss of species richness emphasize the importance of low-intensity habitats in supporting farmland populations. If current trends continue, farmland bird species will continue to decline in Poland. Although there are more studies that need to be conducted, this case study strongly suggests that other Member States of the EU may also be experiencing biodiversity loss. In order to prevent further loss, agricultural management needs to be improved to protect existing species.

Climate Change Effects on Agriculture in Arochukwu

The increasing amount of greenhouse gas emissions significantly impacts agricultural production. As the climate changes, farming seasons become ambiguous, there is rainfall change, extreme weather, and increasing disease; all of these factors can lead to crop failure. By 2100 it is predicted that Nigeria may lose up to 4% of its Gross Domestic Product in agriculture. In order to assess climate change and adaptive measures on crop production in the Arochukwu Local Government Area of Abia State Nigeria, Foster (2013) took a random sampling of 120 farmers. He gave each participant a survey, collecting data on socio-economic status, awareness of climate change, effects of climate change on agricultural products and adaptation. Foster hypothesized that there would be "no significant relationship between the socio-economic characteristics of the farmers and climatic factors affecting crop output." His study found that climate change had a positive impact on most food crops and a negative effect on agroforestry practices. Additionally, farming experience, education and size of the farm had a positive relationship to crop output while age and rainfall had a negative relationship to output.

Foster's study was executed in the Arochukwu Local Government Area of Abia State Nigeria, which is in a tropical zone with an annual rainy and dry season. Foster took a simple random sampling of farmers in 19 blocks of the study area. Eight blocks were selected for the study from which 15 farmers were randomly chosen for a total of 120 participants. Foster administered a pre-tested questionnaire to each of the farmers inquiring about socio-economic characteristics, awareness of climate change, impacts on agricultural products, and adaptation measures employed. During analysis, frequency counts and percentages were applied to descriptive statistics and a simple linear regression to inferential statistics.

In this study the majority (54%) of the participants were female and 35% were between the ages of 20–30. A majority had secondary education (67%), a household size between 6–10 (45%) and were married (57%). For farmers, a majority (38.3%) had a farm between: .51–1 hectares and 11–15 years of farming experience (30.3%). These results indicate that more female-headed households had a farm and a majority of the farmers were literate with the capability to observe and adapt to climate change. In addition, most households had a good number of laborers and were small-scale farmers with a lot of experience. When asked about climate change participants responded that heavy rainfall, temperature increase and erosion were the factors that were prominently noticed; 26.7% observed climate change 6–7 years ago while a minority (15%) only noticed change 2–3 years ago.

To assess the effects of the climate on agricultural production, Foster asked about specific products in three different categories (food crop, livestock and agroforestry). The results showed that climate change had a positive effect on yam, cassava, vegetables, and sweet potato while having a negative effect on food crop labeled "others." Furthermore, there were positive impacts on sheep and poultry but a reported negative effect on pigs, cattle, and rabbits. Unfortunately, climate change had purely negative impacts on all the variables listed in the agroforestry category (snailery, aquaculture, agriculture, mushrooms, and forage). In order to adapt to consequences of climate change 12.9% of farmers reported engaging in mixed cropping, 11.9% in mono cropping and 10.7% in mulching.

The results of the statistical analysis of both the descriptive and inferential data disproved the author's hypothesis that there is no connection between socio-economic characteristics of participants and climatic factors affecting output. The study found that the greater the age, the lesser the crop output. Higher education led to a more effective output. For farming variables the more experience a farmer had, the greater the output, and a larger farm led to a more productive yield. It is recommended that farmers try more adaptation measures in order to mitigate the impacts of climate change on agricultural production.

Predicting Hydraulic Changes

Changes in precipitation and temperature have a direct impact on crops and water. Jha and Gassman (2013) estimated the impact of future climate change based on meteorological prognostications from ten different global circulation under the IPCC model scenario A1B. The delta change method was applied to predict future hydraulic changes in the Raccoon River Watershed (RRW), which was chosen as a representative model for other watersheds. Results using the Soil and Watershed assessment Tool (SWAT) modeling technique showed a decrease in future snowfall, surface runoff, baseflow and total water yield. In addition, stream flow near the watershed was predicted to annually decline on average by 17 percent creating major cause for concern for future crop and water sustainability.

Jha and Gassman conducted this study on the Raccoon River Watershed (RRW), which covers nearly 9400 km^2 across 17 Iowa counties and drains into the Des Moines River watershed. The majority of the RRW area contains corn and soybeans, which makes it especially vulnerable to stream variables such as nitrogen, phosphorus levels, sediment, and bacteria pollutants. In order to test future climate change impacts on this watershed, the authors divided the region into 112 sub watersheds and applied a Soil and Water Assessment Tool (SWAT) model. The model was driven by daily climate data such as precipitation, temperature, and humidity and showed climate change on hydrology and water quality of the shed. In order to obtain a reasonable prediction of future watershed quality due to climate change, this hydrologic model was combined with global circulation model (GCM) meteorological data. The meteorological data (daily precipitation and maximum and minimum temperature) from 10 GCMs under the A1B scenario (rapid economic growth, population growth, and spread of new technologies) were extracted from the 20th century contemporary climate archives (1961–2000) and the future A1B scenarios (2046–2065). Due to significant differences in soil, landscape and drainage features, the RRW was divided into two regions (North and South) for hydrologic simulation. All GCMs under contemporary and future predictions of precipitation and temperature were analyzed for climate changes. Using the delta change method the percent changes in precipitation and absolute changes in temperature for all models were averaged to obtain a mean change. The results of these simulations from the North and South RRW were then analyzed for future hydrological response and stream flow changes in the watershed.

The ensemble mean of the GCMs is predicted to experience a 0.7% increase in daily precipitation and 2.78°C increase in average temperature in the middle of the 21st century (2046–2065) over the entire watershed. Although there will not be a lot of annual variance in percent precipitation, monthly variation is significant ranging from –11.3% to 19.5%. Temperature is predicted to increase 3.9°C during the summer and about 1.9°C in spring. The SWAT model indicates snowfall will decrease by 22%, surface runoff by 16%, base flow by 18% and total water yield by 17%. Evapotranspiration (ET), on the other hand, is expected to increase by 8%. These results indicate a large decrease in overall water supply. The subwatershed level contained a lot of variation with water yield decreases ranging from 10% to 23% but had an overall average 17% decrease. Furthermore, a decrease in water yield indicates a decreased stream flow with the highest impacts taking place during the summer. Assuming these GCMs are accurate, significant mitigation towards crop and water sustainability needs to occur in the RRW region. Best management practices are one solution to reduce non-point contamination such as nutrient and livestock management. Additional research needs to also be done to determine climate variability and climate change impacts on water supplies, food production and the RRW economy.

Kansas: Why Farmers Grow Biofuel Crops

Agriculture has been criticized for its contribution to greenhouse gas emissions (GHG), but there has not been sufficient research investigating how agriculture can be beneficial in a future with climate change. One beneficial solution is biofuel cultivation, which according to the 2009 Renewable Fuels Standard (RFS) would significantly reduce GHG emissions (White and Selfa, 2013). As one of the most productive agricultural states, Kansas was chosen by White and Selfa as a valuable case study to investigate factors that influence the farmer decision-making process regarding innovations. Prior to the case study, White and Selfa conducted a literature review on previous studies done nationally and internationally. From the data they developed a conceptual model of key elements involved in farmer decisions. With this model in mind White and Selfa conducted their own study in Kansas by interviewing 16 key informants with expertise in agriculture and environmental issues, and 17 farmers. The study concluded that farmers were influential in adopting a new idea such as biofuel cultivation by local environmental conditions, communication through existing social relations, the assurance of farmers continual independence, a contribution to a greater societal good, and whether change was more economically advantageous than previous practices.

Before initiating their own case study, White and Selfa conducted a literature review to learn about farmer behavior. In Sweden several farmers were asked about their opinion on growing energy crops and it was discovered that policies need to consider a farmer's personal values and attitudes towards bioenergy crops rather than assuming decisions are purely based on economic grounds. In a 2007 Tennessee study, farmers with higher education and income were much more likely to grow switchgrass as biofuel. On the other hand, the lack of infrastructure for biomass processing and inadequate policies prevented biofuel crops in one study in the UK In studies in Iowa and Kentucky there were concerns about crop profits and equipment prices. This literature review highlighted common themes that influence farmer decision-making and willingness to grow biofuel crops. From this information White and Selfa were able to create a conceptual model for understanding land use decisions made by farmers based on decision setting, natural environment, the farmer himself, and advantage of new practices.

In April and July of 2010 the authors conducted a series of interviews in Kansas to further examine factors that influence farmer decision-making (especially involving biofuel cultivation). Their research was done in two phases; the first phase involved 16 interviews with key informants with expertise in agriculture and environmental issues such as: state agencies, ethanol production facilities and non-profit groups. The second phase consisted of 17 interviews with a diverse group of farmers from 10 different counties in Kansas. All participants were asked questions within their field of knowledge concerning the impacts of environmental issues on farming and decisions, policy influences and farmer decision-making all in regards to biofuels.

In their analysis, White and Selva found many similarities with their existing conceptual model. When making a decision, farmers expressed concerns with the natural environment. They wanted to assure a continual healthy environment but in general their perception of existing environment health was fairly positive. In addition, local influences and local conflict over federal policy, the information available to them, and whether a new practice was profitable and more advantageous than previous practices were very important components in their decision-making process. Any efforts to incentivize and subsidize new practices through policy would most likely clash with farmer values. In this study, the authors found out that climate change was not a salient concern amongst Kansas farmers. They were much more concerned with annual weather patterns rather than long term climate models. Weather is already so extreme in Kansas that farmers do not associate changing weather with climate change. There is ongoing research on this project in Kansas which will provide more data on the ideas in this study. The question of whether Kansas will be a beneficial biofuels generator is still unknown. Continued research is needed at the local, state and national level in order to seriously consider implementing biofuel cultivation.

Reducing Nitrogen Fertilizer Emissions

In agriculture, synthetic nitrogen (N) fertilizer is critical to sufficient food production, but overuse of these fertilizers leads to air, soil, and water pollution. China is the largest producer and consumer of N fertilizer, which causes about 7 percent of greenhouse gas (GHG) emissions in China. Zhang *et al.* (2013) conducted a study in China to quantitatively evaluate the N-related emissions and investigate mitigation strategies. Data were obtained from various sources for each step of the nitrogen fertilizer process including fossil fuel mining, transport, distribution, fertilizer synthesis, transport, and application. Based on these estimates of total GHG emissions from production and utilization of N fertilizer, the authors identified strategies to reduce emissions by implementing newer technologies and methods. Zhang *et al.* determined that using advanced technologies could cut N fertilizer emissions by 20 to 63%, reducing China's total GHG emissions by 2 to 6%.

In order to understand the impact of N-related fertilizer emissions in China, Zhang *et al.* obtained their data from a variety of sources. Emissions from energy production of the fertilizer were obtained from two previous studies. Related ammonia synthesis data were obtained by the Chinese Nitrogen Fertilizer Industry Associations and helped estimate emissions. Pollution from manufacturing N fertilizer was estimated from the developed products urea, ammonia, nitrate, ammonium, bicarbonate, ammonium chloride, and compound fertilizers. Emissions from transporting energy and products were obtained using data from the National Bureau of Statistics of China. Lastly, post-application field emissions were estimated from field measurements of nitrous oxide (N_2O) emissions. These data were compiled to calculate total emissions from N fertilizer in China from 1980 to 2010 and were projected to 2020 and 2030 under a variety of scenarios.

The results showed that ammonia synthesis produces the largest amount of GHG during N fertilizer production at 5.1 tons of CO_2-equivalent (t CO_2-eq).

Manufacturing N fertilizer products emit 0.9 t CO_2-eq while transportation and distribution emit only 0.1 t CO_2-eq. Application and fossil fuel mining emit a combined 7.3 t CO_2-eq. This evaluation assessed that for every ton of N produced and used in China's cropland, an average of 13.5 t CO_2-eq is emitted.

Based on these results, the authors suggest several mitigation strategies to combat emissions. Ideas include improving natural gas leak recovery, heat-conversion efficiency in power plants, and efficiency of ammonia synthesis. If implemented, these strategies could reduce emissions from the N industry from 8.3 t CO_2-eq to 5.8 t CO_2-eq. To further estimate impacts of such strategies, the authors propose four future scenarios: scenario one—business as usual, scenario two—upgrading technologies by 2020 with increasing N fertilizer use annually, scenario three—upgrading technology while keeping N fertilizer use at the 2010 level, and finally scenario four—upgrading technology with annual reductions of N fertilizer use. The results for each scenario by 2030 are as follows: under scenario one an increase to 564 t CO_2-eq would occur, under scenario two a decrease to 403 t CO_2-eq would occur, under scenario three a decrease to 320 t CO_2-eq would occur, and under scenario four a decrease of 206 t CO_2-eq would occur.

This evaluation demonstrates the magnitude of N fertilizer t CO_2-eq emissions in China and provides possible mitigation strategies. N fertilizer will always be necessary but overuse has caused serious water and air pollution. Cutting back in the manufacturing sector should be easiest since there are significantly fewer N fertilizer manufactures than farmers applying it. The fertilizer-use sector also requires significant reductions since this would decrease overall demand and waste. Currently such reductions are difficult due to low fertilizer costs ignoring external costs, lack of a use-recommendation system for farmers, small farm plots inhibiting technology adaptation, and an abundance of part-time farmers whose main focus is not farming. The authors propose that China alter the N fertilizer policies since they are outdated and their modernization would potentially be the most effective in reductions.

Investigating Approaches to Sustainable Intensification of Agriculture

As the global human population grows exponentially every year, the need for food not only increases, but also the demand for a variety of other resources. Since the majority of agricultural practices are damaging to many of these, there needs to be a solution to increase food yields without adversely affecting the other resources as well. Garnett *et al.* (2013) identify four principles of one such approach: sustainable intensification (SI). These principles facilitate an increase in food production while minimalizing environmental damage. The authors point out the importance of using different approaches and implementing and analyzing them in their appropriate context and location. Additionally, the importance of political goals is addressed. Five areas of policy are explored surrounding SI that involve maximum efficiency and benefits while minimizing cost and damages. Nevertheless, it is concluded that SI is a field in progress and is only one piece of a wider solution towards food security and sustainability. In order to explore SI more closely, policy, efficiency, resilience, consumption, and reduction aspects need to be taken into equal consideration.

Garnett *et al.* investigated SI in a broader perspective in order to investigate how and when SI is best implemented. They first explored production goals and identified that food production needs to be affordable and available but the demand for resource-intensive food needs to be reduced. A need for higher production and yield is inevitable due to policy failures and other detrimental factors, so sustainable measures need to be achieved. There is the problematic factor that conversion of land leads to a higher release of greenhouse gases (GHGs). On the other hand, using land for lower yields release fewer GHGs but require more land. Overall, Garnett *et al.* concluded that the food system requires a new design. Since SI presents a goal, not a defined methodology, different approaches need to be implemented and explored in the appropriate surroundings and context.

Another major component to consider surrounding SI is politics. Land use and biodiversity policies need to incorporate the impacts of agriculture and need for conservation as well as assuring high yields. Different methods can be used but the best solution can only be determined in a defined context. A successful SI requires land share efficiency and a balance between yield and environmental benefits. Other dynamics that cannot be ignored when exploring SI are animal welfare and human nutrition. A wider ethical lens should be considering in viewing SI affairs. There is a limitation to meeting livestock demand so the authors believe that demands need to be reduced and that the most promising approach is to improve crop nutrient value through methods such as genetic modification.

In addition to genetic modification there exist other new technologies that must be implemented for SI success. Technology and financial instruments can be used to improve resilience in the market and extend services to new areas (especially in rural areas) to incentivize SI. Most developing countries and low-income communities lack the funds to maximize yields. SI needs to have an agenda that not only aids poor communities but also improves economic growth while avoiding negative impacts. Overall SI is a new developing system and is only a part of the solution to improve food production and security. The food system needs a radical change but Garnett *et al.* stress that the SI issues at hand must be looked at in relation to policy, resilience, efficiency and consumption.

Conclusions

Agriculture is contributing to and being impacted by climate change and must be addressed in order to sustain the health and welfare of the global population. In order to survive, communities internationally need to both mitigate and adapt to a changing climate. Farming management methods need to become more sustainable and more efficient to prevent converting land to agriculture at a high rate. Many approaches such as sustainable intensification and modernizing technology are worthy solutions but require more research. In fact, most studies require further investigation to capture a full understanding of the impacts and social implications. Ultimately, the cost of inaction or slow action is quite high. As temperatures rise and climate shifts the worldwide economy will be impacted by damaged yields. Not only will populations suffer but the economy itself will be in trouble. Furthermore, farmers need to be educated on climate change in order to understand potential damages. Governments should subsidize costs to farmers in order to promote sustainable practices rather than letting them resort to the older and cheaper damaging methods.

References Cited

Arbuckle Jr., J. G., Morton, L. W., & Hobbs, J., 2013. Farmer beliefs and concerns aboutclimate change and attitudes toward adaptation and mitigation: Evidence from Iowa. Climatic Change, 1–13.

Beck, Jan. 2013. Predicting climate change effects on agriculture from ecological nichemodeling: who profits, who loses? Climatic change 116.2, 177–189.

Foster, Alfonso., 2013. Assessment of the Effects of Climate Change and Its AdaptationMeasures on Agricultural Production. European Journal of Climate Change 10.1, 1–8.

Garnett, T., *et al.* 2013. Sustainable intensification in agriculture: premises and policies. Science341.6141: 33–34.

Jha, M. K., Gassman, P. W., 2013. Changes in hydrology and stream flow as predicted by amodeling experiment forced with climate models. Hydrological Processes. 10.10002/hyp.9836.

Rodrigues, J., Pellizari, V., Mueller, R., Baek, Kyunghwa., Jesus, Ederson., Paula, Fabiana., Mirza, B., Hamaoui, George., Tsai, Siu., Feigl, B., Tiedje, James., Bohannan, B,. Nüsslein, K., 2013. Conversion of the Amazon rainforest to agriculture results in biotic homogenization of soil bacterial communities. Proceedings of the National Academy of Sciences 110, 998–993.

Sanderson, F.J., Kucharz, M., Jobda, M., Donald, P., 2013. Impacts of agricultural intensification

and abandonment on farmland birds in Poland following EU accession. Agriculture, Ecosystems and Environment 168, 16–24.

White, S., Selfa, T., 2013. Shifting Lands: Exploring Kansas Farmers Decision-Making in an Era of Climate Change and Biofuels Production. Environmental Management 51.2, 379–391.

Zhang, Wei-feng., Dou, Zheng-xia., He, Pan., Xiao-Tang, Ju., Powlson, David., Chadwick, Dave., Norse, David., Lu, Yue-Lai., Zhang, Ying., Wu, Liang.,

Chen, Xin-Ping., Cassman, Kenneth G., Zhang, Fu-Suo., 2013. New technologies reduce greenhouse gas emissions from nitrogenous fertilizer in China. Proceedings of the National Academy of Sciences 110, 8375–8380.

11. Genetically Modified Organisms: Public Perceptions

Morgan Beltz

Whether to continue using and expanding the use of genetically modified organisms (GMOs) is an ongoing topic of debate on a national and international level. After twenty years of successful modification techniques, there is still much opposition to and skepticism of GM foods. This chapter looks at the last twenty years of genetic modification and the public perceptions that surround it. The chapter starts with a current GM issue, GM fish, works though testing methods to test if a product contain GM ingredients, and then finishes with different perceptions the public has, ending with a summation of twenty years of GMO research. The overall consensus about GMOs is that the science shows them to be safe, but public perceptions still show concern and sensitivity to risks of using GMOs.

Three Scenario-Based Impacts of Transgenic Fish in the Fishing Industry

The increase in fish demand, due in part to population growth and rising income, is causing concern as to how the fishing industry will respond with more product in a timely manner. One possible solution to the fishing industry's problem is genetically modified (GM) fish. Menozzi *et al.* (2012) discuss future trends of the fishing industry and potential risks that GM fish could bring. The authors have created a qualitative scenario analysis based on literature reviews and interviews with experts in the fishing industry. The data collected from the interviews allowed the authors to create three realistic scenarios of the impact of genetically modified fish in the fishing industry. The three scenarios are 1) no market for GM fish, 2) GM salmon reaches the dinner plate, and 3) GM salmon does not take off in the market. The authors concluded that GM salmon will most likely be a part of the solution to the increase in demand for fish, and it is just a matter of when.

Menozzi *et al.* created a three step process for analyzing the fishing industry before they reached the three most likely scenarios of the impact of GM fish.

First they created an accurate description of the current situation of the fishing industry by collecting information from literature reviews and researching online through the FAO to understand the decline in fish population and the increase in demand. The second step was to identify trends and driving forces in the market that are the most likely to be affected by the introduction GM fish. These were identified as public acceptance, regulatory framework, productivity increase, and the market structure in general.

The final step towards the authors' conclusions was to conduct interviews with experts in the industry from all over the world. Fourteen experts in different positions in the fish industry were given a questionnaire to identify key variables that will impact trends for the next generation. The questionnaire also asked questions pertaining to GM fish in the market and the acceptance the experts thought the fish would receive. These answers were cross-referenced to the trends the authors previously identified in order to find links to the forces driving the industry. After all the data were evaluated, the authors identified the three most plausible scenarios for GM fish in the market, 1) no market for GM fish, 2) GM salmon reaches the dinner plate, and 3) GM salmon does not take off in the market.

In the experts' view, the increasing demand for fish was the most important main trend affecting the industry, and the increasing sea temperature was unimportant. The experts interviewed generally agreed that the trends identified by the authors were considered of equal but lower importance. The experts also identified environmentally friendly brands, fish health management techniques, waste treatment innovations, and breeding program improvements as the most important innovations to the fishing industry; GM salmon commercialization was considered as the least important.

The experts disagreed on when the introduction of GM fish to the industry would occur and what the acceptance would be, however the general consensus was that GM fish still have a long way to go before they could be introduced to the market, if they ever get there. The experts also felt that the public will have a hard time accepting GM fish, but consumer and producer acceptance is likely to be higher in emerging and developing economies. The experts did agree that GM fish will bring new regulations to the market, a market price decline, and an unequal distribution of profits. All of the uncertainty among the experts led the authors to the three scenarios.

In the first scenario there is no market at all for GM fish, due to strong resistance from consumers, retailers, and producers. This scenario would force the industry to increase production without GM fish, leading to new developments to minimize environmental impacts and costs, and to improve fish health management, waste treatment, and breeding technology.

The second scenario has GM fish making it to the dinner table in the near future and being completely accepted in the market. In this scenario the authors see market segmentation occurring by geographical location with the producing countries primarily serving nearby countries, but with a reduction in the market price.

Although production would increase, the authors do not believe that the profits would be equally distributed because complying with the new regulations and technologies will be much easier for larger-scale farmers than for small-scale ones.

In the last scenario GM fish do not take off. They are produced and brought to the market, but face resistance from producers, consumers, and even retailers, and are purchased mostly by low-income consumers.

In any event, the experts predicted that having GM fish in the market will increase regulatory oversight and force the aquaculture industry to innovate breeding programs, technical improvements in pens and cages, and waste treatment techniques. Overall this should reduce the market price for all consumers in the long run.

After the creation of these three scenarios the authors went back to the experts to see which one they believe is most likely to occur. The experts' responses identified the third scenario as slightly more likely, but not by much.

Wild type and Transgenic Hybridization of Atlantic Salmon and Brown Trout

With the possibility of genetically modified salmon being approved for the food industry, there is a growing concern of what would happen if a transgenic salmon breeds with a wild salmon. Oke *et al.* (2013) studied what the effects of breeding transgenic salmon and of cross breeding transgenic salmon with a brown trout would be in the wild. The authors created control and cross breeding scenarios which consisted of six brown trout families, two Atlantic salmon hybrid families (the mother was a transgenic salmon), four brown trout families (the mother was a brown trout), and seven salmon families (five families had transgenic mothers and two had transgenic fathers). The authors conducted this study in two different environments; one to mimic hatchery conditions, and the other, in stream mesocosms, to resemble natural conditions. The results show that the hybrids grew more rapidly than either transgenic salmon or non-transgenic fish in hatchery conditions. In the stream mesocosm the transgenic hybrid salmon and brown trout grew 86% and 87% faster than transgenic salmon. However, wild-type salmon grew at similar rates to wild-type hybrid brown trout and salmon. These results show the importance of not allowing genetically modified fish to breed with the wild population.

Oke *et al.* first examined the fish in hatchery-like conditions. Each fish family was raised in an individual tank and fed commercial dry fish feed for 100 days. The families were separated in order to trace the transgene from the parents into the hybrid offspring. Mortalities were removed daily, and every second one was examined for the presence of the transgene. At the beginning of the experiment, the mass and length of 30 individuals from each family were measured and the fish with transgenic parents were genotyped. After 100 days, fish were resampled for mass and length and genotyped to follow the transgene through breeding. Following hatchery-like early development the fish were put in stream mesocosms for 30 days. The authors looked at sympatric and allopatric treatments to see what the difference of having transgenic hybrid fish in natural conditions is. The sympatric treatments contained salmon and either an Atlantic salmon or brown trout transgenic hybrid, whereas the allopatric treatments only had salmon. The fish were fed by drip feeders with feed of variable shapes and colors to mimic natural conditions. Before and after the experiment fish were sampled for length and mass, and genotyped to map the transgene and its affect throughout the study.

The authors used the polymerase chain reaction (PCR) test on dried fish skin for genetic screening. This allowed the DNA strands to be separated and viewed for presence of the transgene. In the hatchery conditions the authors used a generalized linear model (GLM) to evaluate growth among the different cross breeds and genotypes. This model was also used to determine if mortalities were related to genotypes and fitness. The same model was used for the stream mesocosms, but an analysis of variance (ANOVA) was used to determine if growth rates differed among transgenic and non-transgenic salmon and hybrids.

The PCR analysis showed 43% of the 363 fish tested positive for the transgene. Of the Atlantic salmon hybrids, only 37% were positive for the transgene, significantly lower than the 50% expected. Forty-two percent of brown trout hybrids were positive for the transgene, still lower than the expected amount. In the hatchery conditions the authors found that growth rates differed significantly between the transgenic and non-transgenic fish. In all cases the transgenic fish grew significantly faster. These results also showed that the transgenic hybrids grew faster than the transgenic salmon. The authors also measured the mortality rate among the fish, finding that transgenic and non-transgenic brown trout hybrids had higher mortality rates than all the other crosses. The results also showed that the direction of hybridization affects mortality; transgenic Atlantic salmon (AS) hybrids had higher mortality than the wild-type AS hybrids, whereas wild-type brown trout (BT) hybrids had a higher mortality than transgenic BT hybrids.

The stream mesocosm resulted in growth being reduced all around in the sympatric scenarios. In the sympatric scenarios the wild-type salmon growth rate was reduced by 54% when compared to the allopatric scenario, and the transgenic salmon growth rate was reduced 82% relative to the allopatric scenario. When looking at the sympatric scenarios further, the transgenic hybrids grew 87% faster than transgenic salmon, and the wild-type hybrids and salmon grew at similar rates.

This study suggests that the effect of the transgene on juvenile salmon growth is dependent on the environment. Transgenic fish grew faster in hatchery conditions than in more natural conditions. However, the results also showed that the presence of transgenic hybrids suppresses the growth of wild-type fish. With being about 20% more common in the mesocosm, the transgenic hybrids would suppresses the growth of wild-type fish at a faster rate than anticipated. Although this study only looks at the juvenile stage, it shows the possible implications of transgenic hybrids affecting the natural stock. As the production of transgenic fish continues to develop, the authors note the importance of regulating hybridization to prevent a more rapid decline of natural salmon from occurring.

Using PCR testing to Detect Foods With Unlabeled GM Ingredients

Before 2009 Turkey had minimal regulations of genetically modified foods. In a study containing 26 processed soybean, maize, cotton, and canola products, Arun *et al.* (2013) found that 42.3% were positive for GM ingredients. After regulations went into place, the authors reevaluated the amount of GM ingredients in foods using a polymerase chain reaction (PCR) test. The authors found that of 100 samples only 25 contained GM ingredients, a significant decrease since the 2009 study. After evaluating results before and after GM regulations went into place, the authors conclude the regulations have been sufficient in decreasing the amount of unlabeled GM ingredients in food. These results show the effectiveness of regulating GMOs during importation and that PCR is a sufficient way to test for GMOs in food.

Arun *et al.* collected 100 different processed food samples that contained maize, soy, or both. The authors used certified reference materials consisting of soybean powder and maize powder as negative and positive controls for comparison with the samples. For the mildly processed samples and controls the authors extracted and purified DNA using the Promega Wizard DNA isolation kit. Highly processed samples had DNA extracted using the cetyltrimethyl ammonium bromide method and then purified with the Promega Wizard DNA isolation kit. The extracted DNA from each sample was primed in a PCR for an amplification reaction. This process allowed the authors to target a specific strand of DNA in each sample to study it. Each amplified DNA strand was separated and stained through electrophoresis in gel that had ethidium bromide in order to view the macromolecules in the DNA strands. To account for false negatives the authors analyzed each negative sample further for presence of soy lectin and maize zein sequences which are present in all GM foods. The authors also put a cauliflower mosaic virus (CaMV) (a virus that almost all GM food contains) 35S promoter in control samples to compare to false negative results related to PCR inhibitors.

The authors found that 14 of 43 maize samples and 11 of 57 soy samples tested positive for GM ingredients, 25 percent overall. Sixty-three negative samples were confirmed as true negatives with specific amplification of the lectin and zein sequences. The other 12 negatives were negative for both lectin and zein, suggesting that they had little or no soy or maize DNA present.The authors also found that three of the 14 positive samples contained two different GM strands, not uncommon, as other studies have found. Arun *et al.* did have difficulty extracting DNA from 5 negative lectin soy samples, as have other researchers, but the samples were evaluated again to make sure the negatives were not false.

These results are similar to those found in other studies conducted after new regulations are imposed. Other researchers reported having similar problems with DNA extraction with highly processed foods, having so many negatives, and having negatives that did not have any detectable maize and soy DNA. This leads the authors to believe that the extraction process can be improved. However, the consistency with other studies shows the accuracy of PCR testing and effectiveness for tracing GM foods through to importation.

Although it is positive that the regulations have led to a decrease in unlabeled GM foods, some remain in the market, leading the authors to call for additional regulations and stricter enforcements. The success of this PCR study shows the authors that it is an effective way to monitor the GM content in unlabeled foods.

Decrease GM Crop Research to Preserve Agriculture Biodiversity

The question of whether to emphasize research on genetically modified crops or on increasing agricultural biodiversity as solutions to sustainable agriculture is currently a topic of intense debate in the food industry. Jacobsen *et al.* (2013) collected studies from all over the world to assess the present pros and cons of continuing GM crop research and increasing agricultural biodiversity. After compiling the studies, the authors found there are two obstacles to having sustainable agriculture; 1) the claim that GM crops are vital to secure food production, and 2) the shortage of research funds for agriculture biodiversity in comparison to research funds for GM crops. Evaluating these two obstacles in regard to the pros, cons, and economics of GMOs, the authors conclude that research funding currently available to GM technologies would be better spent financing more efficient breeding techniques to cultivate agriculture biodiversity.

Jacobsen *et al.* claimed that there is burden on the environment from the current agriculture practices and increase in monoculture production, in particular the reduction of biodiversity through soil degradation. The authors recognize that GM crops can help solve this problem through technological advances in gene sequencing to improve adaptive capabilities in the crops. However, they argue that GM crops are financially impossible in many developing countries because large manufacturers have a monopoly on the industry and drive up seed cost to a point that is not affordable for small-scale farmers. For example, the authors have found a correlation between suicides of Indian farmers and the prices of Monsanto seeds which have turned out to be overly expensive and have not had the high yields expected by the farmers, leaving them in debt. The economic burden is high for developing countries because the seeds have lowered the labor needed and increased the price, but have not increased the yield.

In larger more regulated countries, such as Denmark, the authors found that GM crops can be beneficial in lowering production costs; sugar beets saved 80 Euros/hectare over non-GMO sugar beets, potatoes saved 108 Euros/hectare over non-GMO potatoes. Maize was the only GM crop that lost money over its non-GMO counterpart at -5 Euros/hectare. These results show that in a regulated established industry GM crops can save money in labor and production. However, they do not guarantee biodiversity and a sustainable way to produce food.

The authors argue that the only way to increase biodiversity is to study the factors that influence it: genotype, environment, and management. They noted that a study on Australian yields over the past 100 years shows that management was responsible for 50% of yield increases, genotype for 35%, and environment for 15%. The study also showed that farms with more agricultural diversity produced higher yields than monoculture farms thus, they argue, funding should continue to go to research the causes of these higher yields.

The authors conclude that all the evidence shows a need for a greater emphasis on agriculture biodiversity instead of GM crops and technologies. They conclude that GM crops will add to the biodiversity loss instead of helping it and will reduce the nutritional value of the soil and hinder the yield. The authors believe that GM crop research should be a basic foundation to learn from for future applications, but not in the short term to increase the world food production. Instead, research should go towards improved agricultural practices, agricultural biodiversity though breeding techniques, and sustainable production. Developing countries are the target of increased food production and practices, and if they cannot afford the GM seeds and have a profitable yield, than the focus needs to change to other sustainable methods.

Global Perceptions of Genetically Modified Foods

Public perceptions of genetically modified foods are not generally the same in different regions of the globe and can help dictate the availability of GM products. Frewer *et al.* (2013) conduct a systematic meta-analysis of 70 journal articles published all over the world, between the years of 1994 and 2010, to compare risk and benefit perceptions of different global regions. The authors focused on papers including agriculture genetic modification. The papers then went through a coding process to detect the levels of risks and benefits presented. The continent results were compared to the mean values of European participants in 2008. The authors found that North America and Asia have a lower risk perception of GM foods than Europe. North America also has a higher benefit perception of GM foods than Europe, but Asia has a lower benefit perception.

Frewer *et al.* created a set of search terms that included GMO, attitudes, and public in some fashion. These terms were applied to online databases of journal articles. The search initially yielded 1638 papers that were then screened with an exclusion criteria to identify the relevant papers regarding consumer attitudes towards genetically modified foods. Papers were coded and weeded out depending on the references to human attitudes, main focus of the study, and sufficient data. This left 70 papers that were then rescreened with a dichotomous scale to rate the papers on intention, attitude, benefit perception, and risk perceptions related to GMOs. The authors note that one of the limitations to their study is the low number of studies from countries outside of North America and Europe. Africa, South America, and Australia each have fewer than five studies, making their data almost negligible. Asia had an adequate size data pool, but still less than half the size of these from Europe or North America, possibly because the authors only looked at papers written in English.

When broken down into specific GM applications—plants, animals, other applications—individuals are more likely to purchase GM plant products and other applications of GM than GM animals. However, when compared to the baseline of the Europe responses in 2008, North American and Asia are more likely to purchase GM animal products. North America and Asian participants saw more benefits and less risk of GM food products than Europeans, with the benefit and risk perceptions increasing each year. Ethical concerns were higher for North America and Asia, but decreased gradually over time. Overall, participants across the entire study had more positive attitudes towards GM plants and non-specific applications, and North America and Asia had more positive attitudes than Europe on the notion of GM products.

This study presents the interesting perspective that geographical location does make a difference in the public's perception of genetically modified foods. Perceptions change depending on the available information and the research of individuals in the country. The authors believe that this information shows the need for increasing research in less developed countries outside Europe and North America to bring awareness of how GM products can help the food industry. They argue that it is important that the research be expanded going forward because public perception is a large part of approving GM products for human consumption. This study shows the importance of research regarding GMOs; it needs to be continued with better methods and reporting of results. Increased knowledge of the public perception of benefits and risks helps producers understand where they need to increase education base in order to increase production of GM foods.

Importance of Engaging Stakeholders in the Genetic Modification Process

Genetic biocontrol technology is one way of controlling invasive fish species. However, like genetically modified organisms, it is controversial in the eyes of the public. Sharpe (2013) studied the public perceptions of genetic biocontrol of invasive fish by conducting eight focus groups in the Great Lakes and Lake Champlain regions. The focus groups were asked the same set of questions and allowed to voice opinions to discuss as a whole group. The discussions of the focus groups were then analyzed in three phases; sorting individual transcripts into reaction categories, coding the text to see emerging themes, and coding the written responses. The author found three central themes in the focus group discussions: issues of uncertainty, acting cautiously, and the question of balance. Most participants thought research for biocontrol was good, but the actual implementation should be analyzed with very high standards. The participants came up with a wide scope of concerns which the author believes is important for developers and researchers to take into consideration.

Sharpe conducted the focus groups in the Great Lakes and Lake Champlain region because they are both facing invasive species problems. The focus groups ranged from 4 to 16 participants with a total of 61. The participants were all people that would potentially have a stake in genetic biocontrol, such as employees of management agencies, therefore not the general public. However, the author noted that not including the general public was of little concern because no participant had a position of authority, no one has vested interests in the approval or disapproval of technology, and all the discussions were confidential. To begin the discussion all participants were provided with the same background information packet that also included the discussion questions. The packet reviewed the genetic manipulation techniques that could be used and the purpose of each one. After the discussions each participant wrote down three benefits of, and concerns about, using biocontrol techniques. These responses were categorized and used as part of the results. The results were arrived at by transcribing the discussion of each focus group and sorting the reactions into categories, coding the categories to find overlapping themes, coding the participants' written responses, and adding them to the different categories.

The author found that the participants generally had four major categories of initial reactions; science fiction, food and agriculture, concerns about the uncertainty and danger associated with the technology, and public perception of the technology. These categories addressed initial fears and controversies regarding genetic biocontrol, such as problems with consumption of GMOs, costs and consequences, and overcoming the negative public perceptions of GMOs. The participants next discussed potential benefits of genetic biocontrol and those responses landed in three broad categories; development of a potential control of invasive aquatic species, other benefits related to the technology, and concerns about benefits. Overall the participants were able to come up with 156 different benefits of genetic biocontrol. These benefits included being able to control species, having more tools to control species, increased knowledge of the technology that could lead to more innovations, and possible creation of an industry. Lastly the participants listed their concerns with this technology, coming up with 300 concerns that fit into five categories: ecological, related to uncertainty, financial, technological, and regulatory. These concerns covered all fears of transgenes being transferred to non-targeted organisms, negative outcomes and impacts, financial costs, success of the technology, and overlapping regulations.

The author concluded from the group discussions that participants feel that the concerns outweigh the benefits with genetic biocontrol. Concerns were much broader than benefits, falling into 11 categories and 22 subcategories, whereas benefits only had eight categories and three subcategories. Although several issues could be cross-referenced as both benefits and concerns, there was too much uncertainty surrounding the issue for it to be seen as an overall benefit. In conclusion, the participants made recommendations for the producers of genetic biocontrol organisms. The recommendations included doing no harm, engaging many different viewpoints, requiring thorough unbiased testing, have a case-by-case approach, clear reasoning for stopping or going forward, an effective regulatory framework, and transparency at all steps.

From leading these focus groups the author found that stakeholders felt that developers do not necessarily consider the viewpoints of others and want to be included in the process. In general, stakeholders want a more determined path of action that includes knowledge of known benefits and harms. These stakeholders are affected directly by genetic biocontrol so they could potentially contribute valid information and knowledge of the implications of this technology to the scientists. Although this was a small sample, the focus groups came up with a wide variety of concerns that could be legitimate issues to solve in the production of aquatic species for genetic biocontrol.

Modeling Risk Frames for Genetically Modified Foods

The regulation of genetically modified organisms (GMOs) is a constant battle of how to balance science with the consumers' perception of risk. Scientists argue that there is no evidence to prove GM foods cause harm, but opponents reply there is no way to look at the potential future risks involved. Lisa Clark (2013) provides a solution to the governance framework by outlining three potential risk frames to use when regulating GMOs; 1) proof of harm, 2) precautious, and 3) precaution through experience. Currently, the regulation of GMOs and biotechnology is split between proof of harm and precautious risk frames, causing tensions between world governments and the correct way to govern the risks associated with GMOs. Clark concludes that individually the proof of harm and precautious risks frames are good, but focus on different aspects of the regulatory process. Bringing the two frames together to create a precaution through experience risk frame constructs the most solid foundation for a concise regulatory framework across different government bodies.

Clark discusses how the proof of harm risk frame endorses the prevention principle which is the idea of preventing the risk at the source rather than having to mitigate the risk after its effect. This means that in order for the GM food to be restricted there has to be science-based proof of harm. The proof of harm risk frame views science as the most important source of knowledge regarding the safety of GM food, but sometimes overlooks other factors. This risk frame does not necessarily look at consumer concern because it does not consider unknown risks, infinite risks, or risky techniques as reason to ban biotechnology. The proof of harm risk frame focuses on the quality of the food grounded in science-based evidence that there is no risk. This proposal also shows how organizations, such as the World Health Organization, are framing the issue of biotechnology with a science-based approach.

The precautious risk frame is perceived as the opposite to the proof of harm risk frame in how it is structured because it embraces the precautionary principal; erring on the side of caution. This frame is structured around preparing and regulating the future unknown risks by preferring to ban GM foods if there is not full scientific certainty that the product is safe. This gives a sense of transparency in the regulatory process by allowing all types of information to be considered, such as consumer prospective and feelings towards GMOs, not just scientific assessment. Clark discusses how transparency in the regulatory process consists of the labeling discussion and having consumer input from the beginning of the process until the food reaches grocery shelves. This way consumers have a say in what products are approved and what is allowed in their food. This risk frame is more concerned about the process of regulation rather than the end product.

Clark discusses the last risk frame as the more forward thinking because it blends both the proof of harm and precautious risk frames into one concise approach. This risk frame understands the complexity of the relationship between science and policy makers. In the regulatory process consumer concerns and scientific evidence have to be taken into consideration, but it can be difficult knowing what the correct balance is. The precaution through experience risk frame looks at both formal and informal decisions about what is the appropriate level of exposure to the risks of GM products. By reviewing formal and informal decisions it shows that science is not the basis for all decisions, but that consumer perception is important as well because consumers are the ones who will choose to buy the products. The precaution through experience risk frame looks at the entire system of approving GM foods by looking at the evidence, science and socio-economic concerns, and creating transparency between regulatory bodies and consumers.

Clark concludes by noting that the different risk frames have powerful influence over the regulatory process of biotechnology and it is important to integrate scientific and sociologic evidence in regulation. GM food is controversial from a policy standpoint so it is crucial to blend science and consumer views in order to make the correct decision when it comes to regulation. The precaution through experience risk frame is emerging on the governance scene, but is increasingly becoming the best substitute for straight science or straight socioeconomic views in the regulatory process. GM foods are the new up and coming technology that cannot be ignored for the future. Clark gives a reasonable solution for the governance issue with the precaution through experience risk frame to properly balance all views involved in the regulatory process.

Twenty Years of Researching Compositional Changes in GM Crops Show Equivalence to Non-GM Counterparts

Genetically modified organisms have been in production for over 20 years, however, there is a perception that they still are not safe because compositional testing still occurs. Herman and Price (2013) review compositional testing of genetically modified crops over the past 20 years and assess if it is still relevant today. Compositional testing has been used to investigate the unintended effects of putting a gene construct into a plant by focusing on the intended effects and safety of gene insertion. Tissue samples of the crops are analyzed for nutrients and antinutrients, then statistical comparisons are performed between GM and non-GM crops to see if differences are observed. If the difference is outside of the statistical range, than the changes are evaluated to determine if the observed levels are unsafe for consumption. The authors review unintended effects for both GM and non-GM crops and conclude that the money invested in testing for compositional changes is no longer relevant after 20 years of safety with very few rare instances of harmful unintended effects.

Herman and Price focus on the unintended genetic and compositional effects of traditionally bred crops in order to compare them with unintended effects of GM crops. Research over the past 20 years has shown that traditional breeding methods are subject to genetic mutations, deletions, and insertions within the genome. Over 2500 crops have been discovered because of this natural mutagenesis. Although some of the mutations are done on purpose by breeders, others are not, and yet few safety issues have occurred through the process. Traditional breeders tend to not focus on compositional changes because breeders try to improve the traits of their crops. The composition also changes due to the surrounding environment and storage of the crop after harvest. The research over the past 20 years has shown it is not accurate to assess traditional crops' safety by the composition.

The transgene is inserted in order to have a desirable characteristic expressed. The insert is sequenced to determine if it has been inserted correctly and will have the desired effect. Genetically modified crops are also bred to ensure that there is only one insertion site, whereas traditionally bred crops are randomly combined and many mutations occur so there is little knowledge of the changes that are happening. Because of the difference in mutation sites, unintended compositional changes are expected. Crops are also expected to be different than those they are bred from because of being sprayed with herbicide or having a gene inserted for disease resistance. Studying these changes between crops for 20 years has not shown any safety hazards. The question is no longer if the composition is the same, rather is the GM crop as safe.

Over the past 20 years the US has reviewed 148 GM crops and Japan 189; every single one of them was found to be equivalent to their non-GM counterparts. The authors conclude that there is a large amount of evidence to deduce that GM crops are less disruptive in compositions than traditional breeding. Because of this, it is easy to say GM crops are completely safe. The authors advise that the money spent on the studies to determine compositional similarities and safety is better spent in research and development of new GM crops. This conclusion was drawn because of the 20 years of safety researched by scientists and the knowledge that we have an increasing global population that needs to be fed, and GM crops are the easiest way to solve that problem. The authors believe that the compositional analysis is no longer needed and it is time to bring the research and knowledge of GM crops to public sector researchers to expand the horizons.

Conclusions

The technology of genetic modification is continuously growing and improving all the time. GMOs have come a long way from pharmaceuticals, crops, and now possibly animals. However, as the technology keeps growing, so does a negative public perception. Genetically modified fish are on the verge of being the first GM animal approved for human consumption, but the concern is the environmental and economic impact. There could be grave consequences if a GM fish escapes and mates with a wild-type fish. However, testing for GMOs is getting better and as a whole, public perception is increasing world-wide a little bit each year. There is still a long way to go for GMOs to be accepted, but risk models and including stakeholders in discussions help the ongoing GMO debate.

References cited

Arun, O., Yilmaz, F., Muratoglu, K., 2013. PCR detection of genetically modified maize and soy in mildly and highly processed foods. Food Control 32, 525–531.

Clark, L., 2013. Framing the uncertainty of risk: models of governance for genetically modified foods. Science and Public Policy 40, 467–478.

Frewer, L., Van de Lans, I., Fischer, A., Reinders, M., Menozzi, D., Zhang, X., Van der Berg, I., Zimmermann, K., 2013. Public perceptions of agri-food applications of genetic modification–A systematic review and meta-analysis. Trends in Food Science and Technology 30, 142–152.

Herman, R., Price, W., 2013. Unintended compositional changes in genetically modified (GM) crops: 20 years of research. Journal of Agricultural and Food Chemistry, doi: 10.1021/jf400135r.

Jacobsen, S., Sorensen, M., Pederson, S., Weiner, J., 2013. Feeding the world: genetically modified crops versus agricultural biodiversity. Agronomy for Sustainable Development 33(4), 651–662.

Menozzi, D., Mora, C., & Merigo, A., 2012. Genetically modified salmon for dinner? Transgenic salmon marketing scenarios. AgBioForum 15, 276–293.

Oke, K., Westley, P., Moreau, D., Fleming, I., 2013. Hybridization between genetically modified Atlantic salmon and wild brown trout reveals novel ecological interactions. Proceedings of the Royal Society 280:20131047; doi:10.1098.

Sharpe, L., 2013. Public perspectives on genetic biocontrol technologies for controlling invasive fish. Biological Invasions, doi: 10.1007/s10530-013-0545-5.

12. Large-Scale Agricultural Pesticides: Unintended Effects and Solutions

Kahea Kanuha

In 1962, Rachel Carson's Seminal work 'Silent Spring' was published. It brought to public awareness the detrimental effects of reckless pesticide use on the environment, particularly on birds. Before its publication, there was virtually no environmental movement in America—agrichemical pesticides were used in abundance without regard to their health and environmental impacts. Since then, some of the most harmful and persistent pesticides have been banned, including aldrin, dieldrin, and DDT, and we have continued to learn more about the unwanted effects of chemical pesticides and alternative methods of pest control.

Pesticides are of an increasing concern because they often have unintentional impacts on non-target organisms and may accumulate in the tissue of animals and humans. Aside from the obvious issue of direct consumption of fruits, vegetables, and grains contaminated with pesticides, they may leak into and contaminate the surrounding soil and water, thus influencing not just the agricultural site of application but the surrounding ecosystems as well. In addition, they persist in the environment for a long time and may continue to have environmental impacts long after initial application. In this chapter we look at recently published studies on the effects of pesticides on various animals, the causes of work-related pesticide poisoning, the use of artificial wetlands as a pesticide remediation strategy, and ways to decrease pesticide use.

We start by reviewing the harmful impacts of pesticides on various non-target animals. Pesticide poisoning has been shown to have a slew of detrimental effects, including reduced neuropsychological performance and information processing in humans, inhibited growth and immune system response in frogs, and DNA damage in rats. We also examine the effects of pesticide contamination on species richness of stream macroinvertebrates.

We then take a look at the rate of pesticide poisoning in humans. Specifically, among agricultural workers, who are often directly exposed to pesticides and are consequently at high risk of pesticide poisoning. Different application techniques and safety procedures may reduce the risk of pesticide poisoning among workers, although they do not reduce the environmental hazards of pesticide use.

One of the major problems with pesticides is that many of them persist in the environment for a long time, often in places where they weren't even applied in the first place. Wind may spread pesticides from the site of their application to other fields. Runoff, particularly soon after pesticide application and heavy rainfall events, brings pesticides to aquatic environments, where they may alter ecosystem structure and function. Remediation strategies have thus been developed to reduce pesticide concentrations in non-target areas. These strategies include capping or blending the contaminated soil with clean soil, incineration, vitrification, and the application or deliberate modification of biochemical processes (Singhvi *et al.* 1994). We will look at the configurations of two artificial wetlands systems as remediation strategies and the costs and benefits of both.

The more we use pesticides, the less effective they become. The chemical warfare raged against rodents, insects, fungi, weeds, and other pests is losing its power. Pesticides do not kill one hundred percent of pests—and as individuals with a higher pesticide tolerance survive and breed, pest populations build a resistance to pesticides over time. More and different pesticides are often the solution to pesticide resistance, yet the more pesticides we use, the more resistant the pests become and the cycle continues, with humans paying more and more each time. Thus, to close the chapter we look at two ways to reduce pesticide use in the first place.

One alternative to increased pesticide use is biological control, the deliberate modification of ecosystems to improve the quality and efficiency of the natural services provided by natural pest controllers. In the United States, the annual value of biological control services is estimated from $4.5–17 billion (Wyckhuys *et al.* 2013) yet the implementation of biological control is often under-implemented and little-understood.

Another alternative to conventional, pesticide-intense farming is the use of different farm management strategies. One such strategy is organic farming, which completely avoids the use of synthetic pesticides, and another is integrated farming, which attempts to find a balance between the economic benefits of pesticides and their environmental costs.

It is unlikely that pesticide use will completely stop in the future—if anything, use will increase to cope with global climate change and increased food demand. For now, the best we can do is to understand more about the unintended effects of pesticide application and try to develop alternatives to agrichemical pesticides.

Pesticide Exposure Reflects Changes in Cognitive Function

Pesticides are widely used toxins, yet research is only beginning to understand their effects on human brain and body functioning. In the United States, it is estimated that close to 8 billion dollars is spent on pesticides each year, yet only a few studies have examined the cognitive and neuropsychological impact of pesticide exposure, to mixed results. Schultz and Ferraro (2013) compared neuropsychological test performance of individuals with an occupational history of pesticide exposure to individuals with no such exposure history. The results suggested that occupational exposure to pesticides results in significant, and age-related, decline in some aspects of neuropsychological performance and information processing.

This study was conducted in the Red River Valley in North Dakota and Minnesota, where mainly barley, durum, spring wheat, edible beans, and canola are farmed in over 4,000 farms across over 4.8 million acres. Participants in the study were prescreened and placed into one of two groups: farmers or nonfarmers. Farmers were defined as individuals with a history of chronic pesticide exposure (defined in this study as pesticide exposure on three consecutive work days) and having performed farm or farm-related work for 1 week of the previous month. Nonfarmers were defined as the individuals who had never performed farm work or been acutely exposed to pesticides.

Each participant went through a battery of neuropsychological tests and surveys, including a demographic survey, the Geriatric Depression Scale—Short form, the State—Trait Anxiety Inventory, selected subtests from the Wechsler Adult Intelligence Scale (third edition), the Mini-Mental Status Exam (MMSE), Boston Naming Test (Short Form), and selected subtests of the Wechsler Memory Scale (Third Edition).

The authors correlated number of years worked with pesticides with the various demographic and neuropsychological measures for the farmer group. Individuals occupationally exposed to pesticides over the course of a number of years developed some aspects of mild cognitive dysfunction. Results showed positive and significant correlations with scores on the MMSE, Logical Memory 1 Recall Total, Logical Memory 1 Thematic Unit Total, Logical Memory 2 Recall Total, and Logical Memory 2 Thematic Unit Total. There was a negative and significant correlation with the State Anxiety raw score, a measurement of present anxiety level. Other correlations were not significant.

There was a significant negative correlation between age and MMSE performance among the farmers, but no significant correlation between age and MMSE performance among the nonfarmers, indicating that cognitive ability declines faster with age in the sample of farmers as compared to the sample of nonfarmers although the two groups were not significantly different in age or other current demographic and psychological variables. MMSE is one of the most used measures of cognitive ability and is consistent with other studies performed in this area.

The authors do not claim that pesticide exposure *caused* cognitive decline, but their results do suggest that pesticide use and exposure may reflect subtle changes in brain function among those exposed to pesticides.

Inhibition of Amphibian Growth and Immune Response by Pesticides

The balance of natural ecosystems is almost certainly being disturbed by human modification of the natural environment. One such modification is the use of pesticides, which may leak into the surrounding soil and water and influence the surrounding ecosystems. Prior lab results have shown that pesticides could affect the health status of amphibians. To verify if similar results would be seen in natural populations, Cristin *et al.* (2013) sampled leopard frogs at three contaminated sites and two reference sites and used size, weight, spleen cellularity, phagocytic activity, and other measurements to examine the differences between frogs collected from contaminated sites and reference sites. The results of the study show that the juvenile frogs exposed to agripesticides are smaller in weight and shorter in length. It was also found that the number of active phagocytes, cells that protect the body by ingesting potentially harmful foreign particles and bacteria, was significantly reduced in frogs exposed to pesticides. Thus pesticide use may inhibit amphibian growth and immune system response.

Cristin *et al.* began by selecting five study sites: three contaminated sites and two reference sites along the tributaries of the St. Lawrence River in Monteregie, Quebec, Canada. The contaminated sites were directly adjacent to agricultural land and were therefore exposed to pesticide runoff. The reference sites included a conservation wetland and a wetland within a rural park, with managed landscape and human activity nearby. Pesticide concentrations were measured through collections of surface water samples. Repeated sampling was necessary since pesticide concentrations may fluctuate considerably with rain patterns and pesticide application time in the surrounding crop fields. For each site, temperature, conductivity, pH, and nitrate concentrations were recorded.

Juvenile *R. pipiens* were captured at the end of July and the beginning of September at each of the five sites. Overall health of the frogs was quantified through measurements of length and weight. The body index, or ratio of weight to length, was used rather than the measurements themselves in subsequent analyses.

To determine immune response impairment, the authors examined the viability of splenocytes, cellularity, and phagocytosis. Splenocytes are white blood cells produced in the spleen, and play an important role in immune system response. The spleen of each animal in the study was removed and cell suspensions were then prepared. Cellularity is the number and type of cells present in a given tissue. Viability, essentially the amount of living cells, was measured to ensure cell extraction methods were effective and that there were enough living cells to proceed with phagocytosis analysis. Both cellularity and splenocyte viability were determined microscopically after trypan blue dye exclusion, which stains dead cells blue.

Phagocytes are cells that ingest harmful foreign particles to protect the body and constitute the first line of defense against infectious microorganisms. Phagocytic activity of splenocytes was determined by introducing fluorescent bacteria to the spleen cells, incubating the cells, and then measuring the fluorescence emission. Higher fluorescence emissions indicate higher bacteria count and thus less phagocytic activity.

The results indicate that frogs living in sites exposed to agricultural runoff are smaller in length and weight. In both July and September, frogs living in agricultural regions had significantly lower body indices than frogs sampled in reference sites. In addition, there were no significant differences of body indices between the two reference sites. Spleen cellularity results show significant decreases in the number of splenocytes for frogs sampled in the three contaminated sites. However, in September one of the reference sites showed a number of splenocytes six times higher than the other reference site. Viability for all groups of frogs captured was over 70%, indicating the splenocyte suspensions were suitable to be used in the phagocytosis assay because there were enough living cells. The phagocytosis results for frogs sampled in July show a significant decrease in phagocyte numbers for four of the five contaminated sites. The September results show a significant decrease in two of the contaminated sites. In both months there was no significant difference between the reference sites.

It has already been observed that pollutants can reduce tadpole growth, but the exact way in which contaminants impair growth is not fully understood. Contaminated environments may hasten a tadpole's metamorphosis in order to escape from a poor growing environment, which leads to a smaller size. However, the smaller size of frogs sampled from contaminated sites in this study may be due to other reasons, such as physiological effects of contaminants on growth or food limitation due to low food quality. The specific interactions between pesticides and the immune system also warrant further investigation in order to fully understand the impact of agricultural practices on aquatic systems.

This study has shown that juvenile frogs exposed to agripesticides are smaller and have compromised immune systems compared to frogs captured in reference sites. Further studies on almost all aspects of the relationship between agrochemicals and ecosystem response are necessary to more fully understand the complex consequences of human actions.

Kahea Kanuha

Organophosphate Pesticides Cause DNA Damage in Rats

The growing use of pesticides in large-scale agricultural applications as well as for household purposes has resulted in their widespread distribution in the environment. To see if pesticide exposure damaged DNA, Ojha *et al.* (2011) evaluated the genotoxicity of chlorpyrifos (CPF), methyl parathion (MPT), and malathion (MLT), three organophosphate pesticides, when given individually or in combination to rats. The results showed that even a single dose of CPF, MPT, or MLT caused significantly high levels of DNA damage in all the rat tissues examined. DNA damage was also observed in microscopic examinations of tissue samples from the liver, brain, kidney, and spleen of rats exposed to one or more pesticides. It was also observed that the damaged DNA is repaired by the endogenous repair systems with time. When the pesticides are given together, they do not potentiate the effects of each other.

Organophosphate (OP) pesticides are a class of pesticide made of esters of phosphoric acid. Many of these pesticides irreversibly inactivate the acetylcholinesterase enzyme (AChE), which is essential to nerve function in insects, humans, and many other animals. AChE inhibition results in uncontrolled neuron firing, resulting in loss of respiratory control. In some cases, this results in death by asphyxiation. OP pesticides degrade much more rapidly than organochloride pesticides such as DDT, aldrin, and dieldrin, which often bioaccumulate since they remain in the environment for a longer time. However, OP pesticides generally have a greater acute toxicity, a measurement of the adverse effects of a substance resulting from a single exposure or multiple exposures in a short period of time.

To study the toxicity of OP pesticides, Ojha *et al.* measured DNA damage in the tissues of rats exposed to CPF, MPT, and MLT, singly or in combination. The pesticides were dissolved in corn oil and given to the rats orally in concentrations based on body weight. The more an animal weighed, the more pesticides given. Rats were killed 24, 48, and 72 hours after exposure. To determine the impact of chronic pesticide exposure, a second group of rats was given the median lethal dose of each pesticide for 60 days.

DNA damage was measured by electrophoresis, which separates DNA fragments based on their size and charge and results in the cell nuclei being visualized on a slide. DNA damage was quantified by the type of "comet" visualized on the slide: intact DNA appeared as round, condensed circles, while damaged DNA appeared as looser, fuzzier blobs. The looser and more scattered the comet, the more damaged the DNA. The results clearly showed that OP pesticide exposure, whether to single or multiple pesticides, caused significantly marked DNA damage in all of the rat tissues examined. There was between a 6.8- and 9.4-fold increase in DNA damage observed in all pesticide-exposed rats when compared to the control rats. Among the three pesticides tested, MPT showed the highest level of DNA damage in all the tissues examined.

When DNA damage was measured 48 and 72 hours after the pesticide treatment, the damage index was lower. This suggests that there is a time-dependent repair of the damage from pesticide exposure, although there were no further tests to see if the damaged DNA would be completely repaired with enough time. The differences in DNA damage among the organs could be explained by the fact that the DNA repair genes are expressed differently in various tissues.

In addition to measuring DNA damage through electrophoresis, Ojha *et al.* examined microscopic changes in tissue samples taken from the liver, brain, kidney, and spleen. Histological changes were seen in the tissues of all rats exposed to pesticides. The authors suggested that these changes might be a result of reactive oxygen species (ROS) causing damage to various membrane components of the cell. Cell structure damage due to ROS is known as oxidative stress and several investigations have examined the correlation between toxicant-induced oxidative stress and DNA damage.

Agricultural Pesticide Use Decreases Species Richness in Stream Invertebrates

Despite years of scientific studies to guide regulation efforts, it is unknown to what degree agricultural pesticides cause species losses and thus a decrease in total biodiversity. To examine this relationship, Beketov *et al.* (2013) analyzed the effects of pesticides on taxa richness of stream invertebrates, using additional analyses to discriminate the possible confounding factors from the effects of pesticides. In all study sites it was shown that pesticide use caused a loss in species and family richness, indicating that pesticide use should be considered a main driver of regional biodiversity loss.

Beketov *et al.* applied the contaminant category richness to study the impacts of pesticide use on freshwater stream invertebrates. Contaminant category richness describes the taxa richness of stream invertebrates particular to different water contaminant levels, quantified by the relationship between the number of individuals and the number of taxa recorded. Data were used from Europe (33 sites) and Australia (18 sites) and included pesticide exposure assessment, stream invertebrate records, and data on environmental factors that may compound the effects of pesticides.

Study sites were assigned a contamination category based on toxic units (TUs) computed from the maximum pesticide concentrations at peak water runoff events. The contamination categories were then compared to the species and family richness of each respective study site. The Chao 2 richness estimator used the collected data to predict the species richness for an infinite number of samples taken at each site. Results show significant differences in taxonomic richness among all contamination categories. The percentage decrease in taxonomic richness between the uncontaminated and highly contaminated sites ranged from 27%, from the Australian family-level data, to 42%, for the European species-level data. Such an extensive decline is comparable to the effects of other known drivers of biodiversity loss. This indicates that for the goal of slowing biodiversity loss, safer standards should be set for pesticide use, methods, and mitigation practices.

Two methods were used to discern the possible effects of confounding factors from the effects of pesticide use. In the first, the SPEAR approach was used to identify the taxa that are particularly vulnerable to pesticides. After these taxa were identified, the data were checked to see whether the overall declines in taxa richness were based on the losses of taxa particularly vulnerable to pesticide use. As expected, it was found that pesticide contamination was associated with a decrease in the number of pesticide-vulnerable taxa. Secondly, available water quality and habitat variables were analyzed to see if any of them would explain the differences between contamination categories. The only significant difference found was in electrical conductivity between the slightly and heavily contaminated sites in Australia, but it was not a consistent linear trend and thus it is unlikely to be a major driver of the observed diversity patterns.

The effects of pesticide use on the taxa richness in Europe were identified in the contaminant concentration level that is considered to be protective by current European agricultural pesticides regulation. This indicates that current standards are not high enough to protect stream biodiversity. Agricultural pesticide use is predicted to increase in the coming decades due to the effects of climate change, and without more stringent regulation it may become an even more important driver of biodiversity loss in the future.

Causes of Work-Related Acute Pesticide Poisoning

Occupational pesticide poisoning is a big health problem among agricultural workers, but there have been few studies on the differences, if any, in risk factors related to severity of pesticide poisoning. Kim *et al.* (2013) interviewed 1,958 male farmers in South Korea to explore work-related risk factors related to acute occupational pesticide poisoning according to the severity of the poisoning. It was found that the risk of acute occupational pesticide poisoning increased with total days of pesticide application, working on larger farms, not wearing personal protective equipment such as gloves or masks, not following pesticide label instructions, applying the pesticide in full sun, and applying the pesticide upwind.

A nationwide sampling survey of male farmers in South Korea was conducted in 2011 by Kim *et al.* who interviewed 1958 households. Twenty-one symptoms of acute pesticide poisoning were selected. Acute occupational pesticide poisoning was defined as a respondent considering himself to have suffered any of these symptoms within 48 hours of pesticide use in 2010. Severe cases of pesticide poisoning were defined as those including symptoms of paralysis or loss of consciousness, while moderate cases were defined as symptoms including vomiting, diarrhea, blurred vision, loss of sensation, slurred speech, and chest pain. The remaining cases were classified as mild.

First the demographics between poisoned and non-poisoned male farmers were compared. Younger age groups had a higher risk of pesticide poisoning, which was possibly because younger farmers engage in more outdoor work than do older farmers. Risk of poisoning increased with annual income and education level. A larger annual income may imply large farm size, which would thus require more frequent or larger amounts of pesticide application. Other demographic factors showed no statistically significant association with acute occupational pesticide poisoning.

The severity of poisoning was evaluated according to symptoms, types of treatment, and number of pesticide poisoning incidents per individual. According to symptom severity, lifetime days of pesticide application, working on a larger farm, and not wearing personal protective equipment such as gloves, pants, boots, and masks were significantly associated with both mild and moderate/severe poisoning groups. The increased risk of poisoning by farm size was likely due to more frequent applications and longer application times. The hands have been reported to be the body part most exposed to pesticides in most cases and highly volatile pesticides are easily inhaled by workers, making the use of gloves and masks especially important when applying pesticides.

"Inappropriate safety behavior" associated with mild and moderate/severe poisoning groups included not following label instructions and applying pesticides upwind. Unsafe behaviors that did not have a statistically significant correlation with pesticide poisoning included not wearing masks and gloves when mixing pesticides, applying pesticides in full sun, not changing clothes immediately after pesticide application, and not bathing with soap after pesticide application.

Previously, orchardists and greenhouse farmers were reported to have an increased risk of pesticide poisoning in South Korea. For this study, interviewees were split into five types of farming: rice, vegetable, greenhouse, fruit, and mixed and other. There was no significant acute pesticide poisoning risk difference by type of farming. The application method was also not significantly related to the risk of occupational pesticide poisoning.

The factors did not show a significant difference by pesticide poisoning severity, suggesting that they may not contribute to pesticide poisoning development in a dose-dependent manner. Prevention strategies for minimizing occupational pesticide poisoning should focus on reducing the number of pesticide applications, enforcing the use of personal protective equipment, and following recommended safety behaviors.

Kahea Kanuha

Artificial Wetlands Reduce Pesticide Concentrations in Surface Waters

As the negative impacts of pesticide use become more and more apparent, there is an increasing need to study pesticide remediation strategies and their effectiveness. Artificial wetlands are one solution to reduce pesticide inputs into surface waters. Tournebize *et al.* (2013) examined two different artificial wetland systems, in-stream and off-stream, to determine the efficiency and effectiveness of each. Both strategies were shown to reduce pesticide concentrations in the water, although each has its strengths and weaknesses. The in-stream option is slightly more efficient and has the added advantage of water storage for re-irrigation of drained lands during times of drought. The off-stream configuration with an open/close strategy shows a similar level of efficiency and requires much less land, making it a good option for farmers to consider. Because of the limitations of both systems, the authors strongly suggest artificial wetlands be incorporated with other strategies for limiting pesticide transfer.

Pesticide fate and transport in natural aquatic environments is difficult to study because there are so many factors involved, including biological degradation, phytoaccumulation, sedimentation, adsorption, and desorption. Along with the many biophysical processes that influence pesticide transport and retention, wetland depth, size, design, location, and inflow variability may also impact retention performance. Controlled experimental conditions have shown that wetlands are highly efficient at reducing pesticide concentrations, but few studies have assessed artificial wetlands in real conditions. Tournebize *et al.* compared two strategies of artificial wetland placements between two similar catchments using water samples collected at the catchment and artificial wetland outlets.

The first artificial wetland placement strategy is in-stream. This strategy treats all water discharged from the tile outlet by detaining the outlet water through artificial wetlands. However, this wetland placement strategy is often not feasible as it requires a large amount of surface area and there is often limited land availability in agriculture-intensive areas. This type of artificial wetland is also unable to fully treat drainage water during runoff events, which is particularly problematic because rainfall-runoff events following pesticide application carry the greatest risk of transferring high levels of pesticides to runoff water. The off-stream strategy involves a manually operated gate to divert tile flows to a series of wetlands during a one-month period following pesticide application, and to close the wetland inlet during times with drainage flow but no pesticide application.

The two research sites selected for monitoring have similar soil and landscape characteristics but different artificial wetland strategies. The results show that in the off-stream wetland, the open/close strategy leads to intercepting only part of the flows theoretically targeted for periods of greater risk of pesticide transfer. Data from the drainage outlet show agrichemical loss does not always follow the predicted pattern of highly concentrated export during the first rainfall-runoff events following application. Wet weather and increased overall rainfall can also result in pesticide transfer when the off-stream configuration is not open. This indicates that to optimize effectiveness, in the future the open/close operation should be driven by the volume of drainage after pesticide application rather than the time since previous pesticide application.

Both configurations gave positive results, but each showed its own limitations in this study. Off-stream and in-stream wetland configuration reduced pesticide concentration in the surface water by a mean of 54% and 45% of all chemicals combined for 2 years. Although both configurations reduced pesticide concentration, Tournebize *et al.* recommend incorporating additional strategies for the reduction of pesticide transfer to the environment.

Each configuration has its own benefits and limitations. The off-stream option provided a similar efficiency in pesticide reduction compared with the in-stream option but took less land out of production. In addition, the storage volume is reduced by 273 m^3 per upstream drained hectare compared to the in-stream configuration. A high storage volume may be difficult and more costly to put in place due to topographical restrictions in many places, as the land drained is usually situated on a relatively flat area. However, the increased water storage of in-stream wetlands can provide supplemental irrigation water in times of drought.

Conservation Biological Control in the Developing World

Conservation biological control (CBC) is the deliberate manipulation of agro-ecosystems to improve the pest control action of natural pest controllers such as arthropod predators, parasitoids, and entomb-pathogens in the vegetation in or around farm fields. CBC practices are beneficial because they do not lead to the development of insecticide resistance, provide pest control and outbreak reduction, and do not bring about human health risks. Wyckhuys *et al.* (2013) reviewed CBC advances in developing regions to identify opportunities to promote alternative pest management. It was found that active research is being conducted in this field and critical insights are slowly but surely being generated, highlighting the need for continued funding for alternative pest management.

Records of CBC practices date back to 300 BC, but development and implementation today is limited due to a variety of problems, including inadequate ecological information about natural pests and ways to exploit them and preconceived opinions about the dominance of generalist predators, which are frequently thought of an impediment to effective biological control.

Wyckhuys *et al.* reviewed CBC advances in regions outside of North America, Australia, New Zealand, Japan, and Western Europe. The literature review was limited to records related to arthropod natural enemies and their effect in controlling field populations of agricultural pests. In total, there were 390 CBC-related literature records from 53 different nations and focusing on 53 focal crops, the most-studied being rice, cotton, maize, coffee, citrus, and beans. The studies were divided into seven categories: assessment of the effect of non-prey foods on natural enemy fitness, understanding of associations with alternative hosts, food, and prey items, determination of the role of artificial food supplements, effects of structural habitat manipulation on resident natural enemies, deliberate manipulation of disturbance regimes, transgenic crops and selective breeding effects on natural enemy performance, and CBC at the landscape level.

Many Southeast Asian countries are currently assessing the role of plant-based food sources for natural enemies of several rice pests. Non-prey foods such as nectar, honeydew, or pollen may act as an alternative food source for natural enemies. A high diversity of food sources for natural enemies can increase many of their fitness parameters and consequent effectiveness of biological pest control. Pollen and carbohydrates in particular are gaining recognition as key food sources for natural enemies, but there is much documentation of how floral resources benefit the pest as well as the natural enemy. Overall, food requirement research seems to be used for improving laboratory mass rearing of natural enemies rather than improving their efficacy in field conditions.

In the literature review, there were 51 references to research on identifying alternative hosts, prey, and food items in a variety of cropping systems. Identification of alternative food or host resources is necessary for the research process, but long-term focused studies are also needed to help guide the definition of CBC schemes and their proper incorporation in agricultural systems.

Nineteen references evaluated the effect of artificial food supplements on natural enemy abundance and efficacy. The food supplements varied greatly and included both carbohydrate sources, such as molasses, honey, and sucrose solution, and protein sources such as pollen, milk, powdered fish, and intestines.

There were 71 records of habitat manipulation and 80 studies specifically on the effect of inter- and cover-cropping on natural enemy abundance or efficacy. The most common habitat manipulation tactics were the use of herbaceous strips, tolerance of weeds, judicious selection of border plants, and the establishment of shade trees. Inter- and cover-cropping studies investigated 54 different plant species. Many of these studies evaluated the effect of a given practice on natural enemy abundance or pest pressure. No records were found in which the underlying drivers for increased natural enemy abundance were determined.

Seventy-seven references were found that described the effects of disturbances such as pesticide use, tillage, and fertilizer application on arthropod natural enemies. Many disturbance factors were taken into account, with the most common being insecticide use frequency or timing, use of organic fertilizer or soil amendments, and alteration in tillage regimes. Most of these factors were evaluated on natural enemy abundance. Several commonly used herbicides and fungicides affect natural enemies, either by directly causing mortality or through modifying habitat structure, composition, or health. Thus, pesticide use reduction can boost the contribution and effectiveness of biological control and save money and resources for farmers.

Twenty-seven studies describe crop-mediated influences on natural enemies. In some crops, leaf trichomes have been shown to facilitate establishment and presence of certain predators while trapping alternate foods such as pollen. Thus, varieties of these crops with enhanced nectar secretion or pubescence is one CBC strategy. Several researchers have identified the potential of breeding crops that resist specific pests and simultaneously encourage their antagonists.

Biological control can be enhanced through an overall increase of landscape diversity or through provision of habitats that benefit one or more key natural enemies. The presence of more stable and heterogeneous habitats such as forests, hedgerows, fallows, or meadows is thought essential to CBC services. They provide resources such as refuge sites, alternative prey, pollen, and nectar for natural enemies. Twenty-nine records were found that describe landscape-level effects on biological control. One example of this is asynchronous rice planting over large areas in Southeast Asia, which allowed natural enemy populations to build up.

The historic lack of financial support and political backing for CBC research and implementation has made the future of CBC seem bleak. However, the documentation of nearly 400 CBC-related literature from 53 different nations and 53 focal crops shows the opposite— active research is being conducted in this field and critical insights are slowly but surely being generated. The benefits of biological control should be recognized, and it should be considered an integral part of sustainable crop intensification strategies in the developing world. This would reduce greenhouse gas emissions, improve human and environmental health, circumvent insecticide resistance development, and promote stability in yield gaps. Funding for further research is necessary to provide much-needed guidance for the development of cost-effective, environmentally sound pest management solutions that are applicable to small-scale farming systems worldwide, especially in developing countries.

Pesticide Concentrations in Stream Sediments from Three Farm Management Approaches

The relative impacts of organic, integrated, and conventional sheep and beef farms on the physical and chemical properties of streams have been studied in the past, as well as the impact on stream macroinvertebrates. To determine if these different farming practices affect pesticide residues in streams, Shahpoury *et al.* (2013) quantified chlorinated pesticides in 100 sediment samples from 15 streams in New Zealand, finding that streams in all three farming categories contained pesticides in the stream sediment, but sediments from conventional farms contained significantly higher concentrations of dieldrin, endosulfans, current-use pesticides, and chlorinated pesticides compared to organic and integrated farms.

There is an increasing global interest in understanding how different farm management approaches affect the environment. Three such management approaches are organic, integrated, and conventional. Organic farming avoids the use of synthetic pesticides and fertilizer, with the goal of using environmentally sustainable approaches to farm management. Conventional farming uses these chemicals. Integrated farming aims to achieve optimal results by finding a balance between the economic benefits of synthetic fertilizers and pesticides and their environmental costs.

Shahpoury *et al.* collected sediment samples from 15 streams passing through sheep/beef farms on the South Island of New Zealand. The farms were arranged in clusters at five locations, each cluster comprising three neighboring farms using different farm management techniques: organic, conventional, and integrated. Sediment sample extracts were analyzed for 19 legacy pesticides and 5 current-use pesticides. The current-use pesticides selected for this analysis have been widely applied, previously detected in groundwater systems, are highly toxic compared to other pesticides, and have high volatilization potentials, making them prone to vapor drift from and between farms.

To more easily understand the data, the pesticide concentrations were split into groups for analysis: dieldrin, endosulfans, current-use pesticides, and all pesticides. Dieldrin was originally developed in the 1940s as an alternative to DDT and was widely used in the 1950s–1970s. However, long-term exposure has proven toxic to a wide range of animals, including humans; dieldrin has been linked to a range of human health problems and is now banned in most of the world. Endosulfans are highly toxic insecticides banned in New Zealand in January 2009. Current-use pesticides are pesticides registered for use in New Zealand when the sampling took place, while legacy pesticides are pesticides, often very toxic, applied in the past that have remained in the environment.

For each of these pesticides or pesticide groups, mean concentrations were found to be highest in stream sediments from conventional farms. A previous study of the same 15 sheep/beef farming streams found that stream macroinvertebrate communities were more degraded in streams passing through conventional farms than through integrated or organic farms. It is certainly possible that the higher pesticide concentrations found in stream sediments from conventional farms have contributed to the macroinvertebrate community degradation.

Chlorinated pesticides were found throughout the study areas regardless of the farm management strategies applied during the 8–11 years preceding the study. One explanation of this is the high volatilization potential of some pesticides. This results in the pesticides being volatized or evaporated into the air and then carried by wind into a different area. As the impacts and concentrations of pesticides are further studied, it seems clear that pesticide use in farming often has unintended and long-range impacts on areas different from those where pesticides were originally applied.

Conclusions

The harmful impacts of pesticide use have become more and more publicized in recent years. Pesticide contamination has been shown to have adverse effects on living beings of all sizes, from the tiniest of stream macroinvertebrates to human beings. These effects may only be expected to be amplified in the future, given the persistence of previously applied pesticides as well as their continued use. Yet there have been steps forward in lessening our dependence on agrichemicals and reducing their impacts. We saw examples of advancements in remediation techniques such as artificial wetlands, and ways to reduce pesticide use in the first place through biological control.

More research is needed on the neurophysiological, cognitive, and ecological impacts of pesticide exposure and use, as well as the development and implementation of remediation strategies. The development of less-harmful alternatives to pesticides will also be crucial in the coming years. Since complete discontinuation of pesticide use is practically impossible given the benefits it provides to agricultural production, the best we can do is understand more about the unseen costs of pesticide use and continue to develop alternatives.

References Cited

Beketov, M., Kefford, B., Schafer, R., Liess, M., 2013. Pesticides reduce regional biodiversity of stream invertebrates. Proceedings of the National Academy of Sciences 110, 11039–11043.

Christin, M.S., Menard, L., Girous, I., Marcogliese, D.J., Ruby, S., Cyr, D., Fournier, M., Brousseau, P. 2012. Effects of agricultural pesticides on the health of *Rana pipiens* frogs sampled from the field. Environmental Science and Pollution Research published ahead of print September 21, 2012, doi: 10.1007

Kim, JH., Kim, J., Cha, E., Ko, Y., Kim, D., Lee, W. 2013. Work-related risk factors by severity for acute pesticide poisoning among male farmers in South Korea. International Journal of Environmental Research and Public Health 10, 1100–1112.

Ojha, A., Yaduvanshi, S., Pant, S., Lomash, V., Srivastava, N. 2011. Evaluation of DNA damage and cytotoxicity induced by three commonly used organophosphate pesticides individually and in mixture, in rat tissues. Wiley Periodicals, Inc. Environmental Toxicology 28, 543–552.

Schultz, C.G. and Ferraro, F.R., 2013. The impact of chronic pesticide exposure on neuropsychological functioning. The Psychological Record 63,175–184.

Shahpoury, B., Hageman, J., Matthaei, C., Magbanua, S., 2013. Chlorinated pesticides in stream sediments from organic, integrated and conventional farms. Environmental Pollution 181, 219–225.

Singhvi, R., Koustas, R.N., Mohn, M, 1994. Contaminants and remedial options at pesticide sites. Risk Reduction Engineering Laboratory, U.S. Environmental Protection Agency.

Tournebize, J., Passeport, E., Chaumont, C., Fesneau, C., Guenne, A., Vincent, B. 2013. Pesticide de-contamination of surface waters as a wetland ecosystem service in agricultural landscapes. Ecological Engineering 56, 51–59.

Wyckhuys, K.A.G., Lu, Y., Morales, H., Vazquez, L.L., Legaspi, J.C., Eliopoulos, P.A., Hernandez, L.M., 2013. Current status and potential of conservation biological control for agriculture in the developing world. Biological Control 65, 152–167.

13. Climate Change, Pesticides, Bees, and Wild Pollinators

Lia Metzger

In the last decade, heightened concern for the decrease in managed honey bee populations and insect pollinators in agricultural fields have given rise to extensive hypotheses and studies. While many of them link the declining population to diseases, many have found strong correlations between the increase in temperature due to climate change and pollinator deaths and decrease in pollinator services. Additionally, neonicotoid pesticides and miticides have recently been related to the decline of bees because they affect their foraging behavior and decrease their motor abilities.

On the other hand, several studies have sought ways to ameliorate the declining population of bees and pollinators. These studies investigate how the biodiversity of the pollinators, the richness of pollinators, the presence of wild pollinators, and mass-flowering crops affect honey bees and other similar insect pollinators. Even though climate change and pesticides have been linked to the decline in population of honey bees in particular, their negative effects have been balanced by wild insect pollinators and changes in their phenologies have buffered them against the increasing temperatures.

The global human population is increasing rapidly, and with it, the demand for major crops is increasing beyond the amount we can produce. One of the alarming parts of the decline in honey bees is that they are commonly used to pollinate mass crops in North America and Europe. Recognizing the extinction of many bee species in these places, and the possibility of more extinctions and less prolific pollination by managed honey bees, many scientists have examined other options for pollinators and the validity of changes in pollination and pollinator populations. In the first paper, the authors investigated the changes in the network of forbs and bees over 120 years in Illinois, USA. The results support the evidence of the negative effects of climate change because hundreds of plant-pollinator interactions were lost, mostly due to bee species extinction and to spatial and temporal incoherence.

Both of these reasons for the decline in the strength of the network reveal the gradual, but detrimental effects of climate change and habitat loss on insect pollinators.

Moreover, the next paper reviews the recent discoveries on the effects of pesticides on honey bees. The authors found that honey bees exposed to acetylcholinesterase inhibitors and neonicotinoids had impaired motor functions and neurophysiological functions and at common levels, died. In the first paper, the honey bees were exposed to four different inhibitors and then observed for changes in behavior, which resulted in increased grooming and flipping over, which could decrease effective foraging behavior and thus, lead to decreased nutrition. The second study specifically investigated the effects of cholinergic pesticides, such as neonicotinoids and miticides that are commonly used, on the neurophysiology of honey bees, and found that the action-potential of the Kenyon cells were significantly affected to reduce neurophysiological function and increase memory and learning loss. Most pollinators are exposed to pesticides either in the air or from the crops, so the third study focused on the exposure of honey bees to neonicotinoids in the clouds emitted from crop machines. The twenty-foot wide clouds were found to cause fatalities in bees when exposed to humidity, and almost all bees had poisonous levels of neonicotinoids.

The last studies focused on how biodiversity of pollinators and other wild pollinators effected the crop production and the support of declining pollinator populations. The authors found that mass crops increased the abundance of solitary bees, contrary to belief, because mass crops such as oilseed rape have a higher concentration of flowers per square foot than wild crops, providing more food for the bees. In addition to mass crops possibly helping rather than harming bee populations, one study found that bee synchrony with apple crops and increased bee biodiversity balanced out the negative effects of climate change. Two other studies investigated how wild pollinator species affected pollination. Regardless of the presence of honey bees, wild pollinators were found to enhance the fruit set of crops more than honey bees and more effectively pollinate. In another study, native bee species buffered the loss of pollination by honey bee species due to climate change in pollinating watermelon. These recent research pieces support more biodiversity of pollinators in order to combat climate change and still increase food production.

Species Loss, Function, and Co-occurrences of Plant-Pollinators Over 120 Years

Over the past decade, bee populations have been decreasing significantly in North America. While many studies have investigated why there has been a decrease, few have researched the long-term change in species richness, in interaction between pollinators and plants, or in function of pollinators. Burkle *et al.* (2013) studied the loss of species of plant-pollinators, focusing on bees, and forbs, their interactions, and the function of bees over 120 years. Using data collected by

Charles Robertson from 1888 to 1891 and data collected in 2009 and 2010 from natural habitats near Carlinville, Illinois, USA, the authors quantified and analyzed the changes in the network structure, bee diversity, and phenologies of bees and forbs. Additionally, data from 1971 to 1972 in Carlinville were used to investigate the changes in bee diversity, quality of pollination, and bee visitation rates to *Claytonica virginica*. Over 120 years, a substantial number of species interactions and bee species were lost and bee phenologies shifted significantly. The authors found that richness of bee species and the rate of visitation to *C. virginica* declined dramatically in the last 40 years and that there was a loss of redundancy in bee species.

In order to analyze the changes in network structure and phenologoies of pollinators and forbs, Burkle *et al.* sampled plant-pollinator interactions and their phenologies in woodlands that were within ten miles of Carlinville during March through May of 2009 and 2010. Twenty-six spring-blooming forb species and 109 pollinating bee species were observed during the peak time of 9:00 to 15:00. The authors then calculated the number of interactions lost and gained between the 1890s and 2009/ 2010. Additionally, to test for the diversity of visitors and the quality of the pollination, the authors used a second historical data set from 1971 from the same fields visited in 2009/2010.

To account for possible sampling differences, the authors used their sampling effort to extrapolate and found their sampling to be close to the "true" richness. Robertson, however, sampled only 13 of the 26 forbs observed in the current study, so the loss of species and interactions and loss of bee diversity is a conservative estimate. Burkle *et al.* also created a null model that used real data and examined the range of possible shifts in bee phenology in order to find how the phenological shifts of bees and forbs contributed to bee extirpation and interaction losses.

Burkle *et al.* found that the overall network structure changed significantly from the late 1880s to 2009/2010. While only 24% of the original interactions were still intact, there were over 100 new interactions between forbs and bees, so only 46% of total interactions were lost. One major contribution to the loss of interactions were bee extirpations, which accounted for 45% of lost interactions (no forbs were lost). More specialists were lost than generalists, possibly because of a higher sensitivity to changes in resources and temperature over time and lower abundance in the 1890s. Parasitic and cavity-nesting species were also lost more significantly than other species, which could be due to decrease of available woody debris and changes in trophic levels. Moreover, 41% of the lost interactions that were a result of bee extirpations were connected to the lack of spatial co-occurrence and/or temporal co-occurrence.

From examining the phenologies of forbs and bees, the authors observed that the peak forb bloom was 9.5 days earlier and the peak bee activity was 11 days earlier. While the phenologies of forb species that peaked earlier versus later in the season did not change, the phenologies of bee species that were active earliest changed the most. The null model showed phenological shifts only could account for up to 55% and 44% of that bee extirpations and loss of interactions. Additionally, the comparision of pollinator diversity and pollination quality between the 1880s, 1971, and 2009/2010 found that the richness of bee species did not change for the first 90 years, but decreased by over half in the past 40 years. Bee visitation to *C. virginica* was four times as high in 1971 as 2009/2010. Bee communities were less intact in 2009/2010 than in 1971, as indicated by a change in significant nesting, in the ecological network sense, of bee species.

Overall, the quality and quantity of bee pollination of *C. virginca* declined dramatically in the last 40 years. While there have been major losses in species interactions, bee species, and pollination service, the increase in different species interactions and the significant phenological shifts reveal a possible resilience to total species lost. However, the considerable loss of specialist bees and the quality and quantity of pollinator service suggest a continued loss of efficient, abundant, and diverse pollinators.

Honeybees Exposed to Acetylcholinesterase Inhibitors exhibit Impaired Motor Function

Stress from parasites, pathogens, and pesticides have been contributing to the global decline of populations of honeybees and many pollinators for the past two decades. Specifically, the use of pesticides that affect neuromuscular functioning and kill parasitic mites have caused the accumulation of acaricides, or mite pesticides, in the wax combs of bees' hives. To investigate the possibility of this accumulation contributing to the decline of bee populations, Williamson *et al.* (2013) studied the effects of prolonged exposure to pesticides that inhibit acetylcholinesterase (AChE) on the physiology and behavior of bees. Adult worker bees were fed sub-lethal concentrations of four AChE inhibitors in sucrose solutions and then were observed for walking, stopped, grooming, and upside down behavior. After the behavioral study, the bees were dissected to confirm that the four compounds they assayed or their metabolites were responsible for the change in behavior by testing for transcript expression levels of two honeybee AChE inhibitors and through biochemical assays. All AChE inhibitors caused increased grooming behavior, but coumaphos in particular caused more grooming and symptoms of sickness as the concentration increased. The authors found that the effects of pesticides that inhibit AChE on the motor functioning of bees could reduce their survival and contribute to the decline of bee colonies.

Numerous studies have focused on the effect of pesticides on bees, with uncertain conclusions as to how chronic exposure to acaricides in hives may affect the behavior of foraging worker bees. Williamson *et al.* chose to experiment on adult foraging worker bees in both the winter and summer with four compounds of AChE insulators in order to consider prolonged exposure to pesticides that affect motor function. The authors compared the effects of 10 nM of the four AChE inhibitors—coumaphoes, chlorphyrifos, aldicarb, and donepezil—in 1 M of sucrose solution with the control of 1 M of sucrose and investigated the effects of three different doses of coumaphos on bees.

Bees were caught, placed in boxes, and fed the different solutions from feeding tubes at liberty for 24 hours and the feeding tubes were weighed before and after the feeding time to determine drug consumption. After feeding, the authors observed each individual bee for 15 minutes for walking, flying, remaining still, falling upside down, grooming the head, and unusual abdominal spasms, noting frequency and time intervals for each movement. To test for IC_{50} values for AChE, the bees were dissected after observation, and subsequently, the Bradford assay was used to determine protein concentrations and the Ellman's assay was used to assay AChE activity. Using semi-quantitative PCR amplification, the authors tested for AChE gene transcript levels in the brain and gut of the honeybee to find what the AChE inhibitors effected the transcript levels. A principal components method of factor analysis was used to find correlations between the behaviors for time intervals. Comparative graphs and tables between winter and summer for each type of AChE inhibitor were created using a multivariate general linear model and least square differences.

The authors found that four factors accounted for 82.8% of the variation in the data, each of which indicated how the bees' expression of behavior was affected by exposure to the AChE inhibitors. Overall, bees did not walk as much, could not right themselves well if they fell over, and had abdominal spasms that the control group did not have. In the summer, the bees treated with chlorpyrifos significantly exhibited less walking behavior, more time upside down, and abdominal spasm compared to the control bees. In the winter, the bees did not show a difference in walking behavior, but in the summer, the bees walked less. In both seasons, there was an increase in grooming behavior. In addition, bees did not exhibit significant differences in their behavior of remaining still or flying with exposure to AChE inhibitors.

The authors further separated the behavior of grooming into the head and body because the AChE inhibitors had the most significant effect on this behavior. The total time spent grooming the head but not the body was increased, and aldicarb and coumaphos contributed to the increased time spent head grooming the most. Both heads and bodies were more frequently groomed in all treatment groups than the control. In particular, for the AChE inhibitor coumaphos, the highest concentration (1μM) treatment group expressed head and body grooming behavior and abdominal spasms more than the controls in the winter but not in the summer.

Biochemical assays found that chlorpyifos oxon and coumaphos oxon, and aldicarb sulfoxide, metabolites of AChE inhibitors, showed greater inhibition of the enzyme AChE than parent compounds. Chlorpyrifos oxon was the most potent inhibitor of AChE for both gut and brain tissues, but the AChE inhibitors were more potent for gut tissues than for those of the brain (lower IC_{50} values in gut). AChE gene transcription also increased in both the brain and the gut for AChE-2 transcripts, mostly in response to coumaphos and aldicarb in the brain tissues and in response to coumaphos and chlorpyrifos in the gut tissues.

The increase in the frequency of grooming behavior, the impaired ability to flip over, and the exhibition of abdominal spasms indicate that AChE inhibiting pesticides alter the behavior of honey bees. Bees exhibited lower levels of AChE in the brain, which could indicate that the behavior of the bees would be affected long-term. It is possible that the increased grooming behavior could be helpful to ridding mites by using coumaphos, but coumaphos also increased abdominal spasms. The spasms could suggest pain, disruption of gut function, or even a gut parasite that has been correlated in high levels with acaricides in combs. Moreover, the metabolites of coumaphos, chlorpyrifos, and aldicarb inhibit AChE in both the brain and gut tissue, which suggests a correlation between the environmental levels of the parent compound with the active metabolite. This correlation could help predict the percentage of AChE inhibitors in comb wax that contaminate the bees. Additionally, the increase of AChE-2 transcripts suggests that long-term contamination may result in increased AChE levels as a heritable trait, which could help the bees adapt to the inhibition of AChE-1 by pesticides. These possibilities, however, do not overshadow the evidence of pesticides affecting the behavior of honey bees, and potentially contributing to the decline of the bees by changes in foraging behavior and gut disruption.

Cholinergic Pesticides Cause Negative Neuronal Effects in Honeybees

Recently, pesticides that target cholinergic neurotransmission have been found to aid in the decline of insect pollinators. In particular, neonicotinoids (nicotinic receptor agonists) and organophosphate miticides (acetylcholinesterase inhibitors) are commonly used and thus, frequently come in contact with honey bees. Palmer *et al.* (2013) investigated how these pesticides affect the neurophysiology of honey bees by using recordings from mushroom body Kenyon cells. Instead of studying the learning and behavior of honey bees that are exposed to neonicotinoids and organophosphate miticides, the authors used whole-cell recordings from Kenyon cells in honey bee brains, and assessed the native connectivity and nAChR expression in KCs. They found that the two neonicotinoids, imidacloprid and clothianidin, and coumaphos oxon decreased the KC excitability by inhibiting action-potential firing and reduced KC responsiveness to ACh. When the honey bee brains were exposed to both neonicotinoids and miticides as is common in large crops, the combined exposure added to the effects on KC excitability and nAChR-mediated responses. The honey bees are usually exposed to much higher concentrations of cholinergic pesticides, which indicates that the negative effects on the neurophysiological responses of the mushroom body cells would be heightened in reality.

In order to study neurophysiological effects of pesticides on honey bees, Palmer *et al.* analyzed the mushroom body Kenyon cells in the honey bee brains for action-potential effects and inhibition of membrane activity. They studied KCs in particular because they are the most significant cells in the honey bee brains for neurological function. In addition, KCs from live tissue were used instead of cultured KCs because cultured KCs do not give full readings of the effects of cholinergic pesticides, according to their image comparisons. Imidacloprid and clothianidin were used because they are the most common neonicotinoids that honey bees are exposed to, and coumaphos oxon was used because it is produced from metabolizing Coumaphos, a common organphosphate miticide that is becoming a frequent subsititute for neonicotinoids.

Specifically, the authors anaesthetized adult worker honey bees, isolated the intact brain, and removed the tissue and membranes to obtain whole-cell recordings from KCs. While taking recordings, the brain was in a recording chamber, where it was secured with a mesh weight and continuously perfused with extracellular solution at room temperature. Recordings of whole-cell voltage-clamp and current-clamp of the mushroom body KCs were taken. The authors used an EPC-10 patch-clamp amplifier controlled by Patchmaster software to record the membrane currents (I_M) and the resting membrane potential (V_M). To find the transient nAChR-mediated responsed, the authors used pressure application of ACh 25–50 μM from the KC. In addition, the Bradford assay was used to determine protein concentrations in the brains and the Ellman assay was used to determine AChE activity. Using appropriate concentrations, the AChE inhibitors were incubated in honeybee brain lysates for 20 minutes, and then samples were incubated with a color-indicator reaction mix and monitored by absorbance for AChE activity. Graphs representing the KC current effects and AChE inhibitor effects on the current and membrane potential were included for both neonicotinoids, coumaphos oxon, and the combined pesticides. Images of KCs from live tissues and comparisons of currents between cultured KCs and live-tissue KCs were also included.

Palmer *et al.* found that neonicotinoids cause rapid, concentration-dependent depolarization of KC membrane potential. Action potential firing under neonicotinoids occurred during the initial development of the depolarization but they did not continue during the plateau phase. Similarly, coumaphos-oxon caused a concentration-dependent depolarization of the membrane potential and a lack of AP firing during the plateau stage, although the depolarization occurred more slowly. As far as the effects on KC ACh responses, neonicotinoids that were applied by batch to the KCs caused a tonic inward current, which indicates a sustained activation and desensitization of KC nAChRs. This means that imidacloprid and clothianidin reduce KC responsiveness to ACh. For coumaphos oxon, AChE activity was initially potentiated, but with continued exposure, the tonic inward current was developed, meaning that AChE was inhibited. With higher doses of coumaphos oxon, potentiation and inhibition occurred more rapidly. Finally, when simultaneously exposing KCs to neonicotinoids and coumaphos oxon, the effects on KC function were additive, causing a sustained depolarization and inhibition of ACh responses.

The effects of these pesticides on KCs indicate that the neurophysiological function of honey bees is severely impaired, causing multisensory integration, associative learning and memory, and spatial orientation to be impaired because these functions are dependent on the mushroom cloud KCs. While many studies have shown that the memory, learning, foraging, and navigation behavior of honey bees have been negatively impacted by neonicotinoids and coumaphos, this study links the behavior to concrete neurological activity.

Mortality of Bees Exposed to Neonicotinoid Clouds around Corn Drilling Machines

Several studies have linked the spring sowing of maize seed to the lethal poisoning of bees and have found that worker bees coming in contact with the exhaust from the drilling machines became contaminated with insecticides and rapidly died. In previous studies with still caged bees and free bees, the correlation between the particles from the drilling machines and the poisoning of bees was not clear. To specify how close the bees have to be to be poisoned, Girolami *et al.* (2013) incorporated the distance from the drilling machines, the type of drilling machine, and the number of times the bees had to pass by the machines before they were killed. The authors also ran the drilling machines with 200 g of talc added to the seed containing hoppers during three trials in order to capture the extent of the exhaust cloud. After being exposed to high humidity, bees that were moved alongside the machines were lethally poisoned. No significant difference in bee mortalities was found between modified and unmodified corn drillers. The exhaust cloud extended approximately 20 feet around the drilling machine.

In order to mimic the movement of foraging bees, ten cages, each containing one bee, were attached to a four meter aluminum bar, held 2.5 meters high, and walked by the side of the drilling machine. In some trials, the bees were held at 2, 4, and 6 meters from the still drilling machine, and then placed in either lab humidity or high humidity to test for the lethality of the neonicotinoid cloud. In addition, the bees were walked 1–5 meters or 5–9 meters from the right side of the mobile drilling machine and parallel to its movement for 30 seconds. Then the caged bees made a U-turn around the drilling machine and were walked along the left side, reflecting the movement of bees making a round-trip around the sowing field. To test for the exposure to insecticides in the exhaust, the bees underwent chemical analysis and were only exposed on one side of the drilling machine in the same way. This allowed for the authors to differentiate the levels of contaminants on each side of the drilling machines and the lethality of the exhaust at varying distances from the machines.

The authors found bees that had been exposed at 2, 4, and 6 meters by passing by the drilling machine rapidly died from clothianidin if they were placed in high humidity afterward. Mapping the deaths of the bees between 2 to 12 meters on the right side and up to 8 meters on the left side of the drilling machine, the extent of the neonicotinoid cloud was measured as approximately 10 meters on each side and 2 meters high. Possible change in the shape of the cloud was considered, but the proposed elliptical shape formed by wind and movement would not reduce the extent of the cloud. Even though the modified drilling machines direct their exhaust at the ground instead of at a 45° angle there was no significant difference between the mortality of bees passing by modified or unmodified drilling machines. Machines sowing seeds that were coated in fungicide did not cause significant poisoning to the bees, but the coatings clothianidin, imidacloprid, and thiamethoxam all caused more than 50 % of the bees that passed by to die. Furthermore, the chemical analysis revealed that all of the bees contained very large quantities of insecticide, with the highest levels of insecticide in bees passing by at 1 meter and decreasing in insecticide levels at greater distances.

The poisonous exhaust can account for the mass deaths of bees in sowing season from corn drillers that are used during this time. Although there were differences in levels of poisons in the bees, the majority of the bees that passed by the drilling machines in the way that bees normally forage were killed. The authors showed that, even with modified drillers or insecticides not affecting the bees directly, the airborne contaminants are extremely lethal to bees.

Wild Insect Pollinators enhance Crop Production irrespective of Honeybees

Recently, the abundance and variety of wild insect pollinators have significantly decreased globally. This poses a threat to the ability of crops to produce enough food for the rapidly growing human population. With fewer wild insect pollinators, crops that rely on animals to spread their pollen will be limited in their ability to reproduce and thus will likely produce a smaller crop yield. European honeybees are often used to assist the pollination of crops, but the research of Garibaldi *et al.* (2013) suggests that this strategy may not be the most effective for food production. Garibaldi *et al.* investigated the amount of pollen the wild insect pollinators and the honeybees deposited on flowers and the mean fruit set of crops around the world to determine if wild insect pollinators and honeybees enhance pollination of flowers and increase the fruit set of crops. The results imply that, while both wild insect pollinators and honeybees increase pollination of crops, the wild insect pollinators pollinate crops more effectively than honeybees and honeybees are not a substitute for wild insect pollinators, but a supplement. In lieu of this, the intentional management of a combination of wild insect pollinators and honeybees may improve global crop yields.

Garibaldi *et al.* studied the pollination and fruit sets of 600 fields in 41 crop systems in 19 countries in all regions of the world except Antarctica. The authors studied pollinator-dependent crops of annual and perennial nuts, seeds, and fruit crops to account for a wide variety of crop types. Crops varied in management practices, landscape styles, abiotic and biotic factors, and native and non-native ranges. Many factors could affect the trends for pollinators, so Garibaldi *et al.* standardized the crop systems from which they collected data. Each of the crop systems consisted of the same species but from at least three spatially separated fields with similar management. This reduced the possibility of random factors significantly affecting the results.

The authors created strict criteria to consistently count the visitations by an assemblage of wild insects to crop flowers in the sampled fields. Pollen deposition was calculated as the number of pollen grains per stigma and fruit set was determined by the percentage of flowers setting mature fruits or seeds. In order to analyze whether the wild insect pollinators and the honeybees enhanced the crop yield, the pollen deposition for each insect species was compared with the resulting mean fruit set of each field. The variation in space and time of pollen deposition and fruit set was accounted for as the coefficient of variation (CV). Descriptive and explanatory graphs with CV for each species of crop and assemblage of wild insect pollinators and honeybees were included.

The study found that crops with more visits from wild insects and honeybees had more pollen on their flowers than crops with fewer visits. Honeybees were responsible for more pollination than wild insects by 74%. This did not match the prediction that crops pollinated by honeybees would have higher mean fruit sets. On the contrary, for all crop systems visited by wild insects, fruit set increased significantly, whereas fruit set only increased in 14% of the crop systems with only honeybee visitation. Disparate visitation and unequal abundance were shown not to be factors that could have affected this trend.

Increased pollinator visitation, however, did not result in as significant an increase in fruit set as the pollen disposition on flowers. The authors considered pollen excess, seed abortion, and filtering of pollen as possible reasons for the visitations not always resulting in pollination. The results suggest that wild insects were more efficient pollinators than honey bees because the difference in pollen disposition and fruit sets was much wider for honeybees than for wild insects. In addition, fruit sets increased in crop systems with visitation by wild insects, whether honeybees visited frequently or not. In crops with both wild insects and honeybees, the fruit sets increased more than crops without honeybees. These results suggest that honeybees supplement the pollination of crops by wild insects but cannot replace it.

The increase in fruit sets with more visitations by wild insect assemblages and honeybees suggests that the integration of managed pollinators with wild pollinators could produce better crop yields. The authors found that wild insects more effectively pollinate crops while honeybees pollinate more with less results. Since there was no negative correlation between the two pollinators when they visited the same crops, integrating the pollinators would likely produce more crop yields. The consideration of wild insects as part of the management of crops could encourage diversity of animal pollinators and increase food production globally.

Mass-flowering Crops Positively Affect Wild Bee Brood Numbers

The expansion of mass-flowering crops has been linked to the loss of biodiversity of farmlands because they escape into natural and semi-natural habitats. However, these mass-flowering crops have a higher density of flowers than non-crop species, and thus produce more food resources with more access to nectar and pollen, so they may enhance the abundance of wild foraging bees. Holzschuh *et al.* (2013) investigated how oilseed rape, a mass-flowering crop, affects the abundance of the solitary and polylectic Red Mason Bee *Osmia bicornis*, a generalist bee species that nests in both natural and semi-natural habitats. Using data from 67 sites in Germany, they compared the abundance of *Osmia bicornic* in grasslands adjacent to oilseed rape fields and isolated from oilseed rape fields and vice versa. Artificial nests were assessed for number of brood cells and for the percentage of oilseed rape pollen in larval food, and then compared between brood cells and the percentage of oilseed rape pollen. The authors found that *Osmia bicornis* colonized artificial nests in grasslands and oilseed rape adjacent to each other significantly more than in grasslands that were isolated from oilseed rape, and not at all in isolated oilseed rape. In addition, in landscape scales, more or less oilseed rape had no effect on the number of brood cells. Oilseed rape pollen in larval food increased in adjacent fields compared to isolated fields.

While the effects of mass-flowering crops on the biodiversity of natural and semi-natural habitats has been well studied, the potential positive relationship of mass-flowering crops and wild bee abundance has hardly been investigated. Knowing this, Holzschuh *et al.* picked oilseed rape as a mass-flowering crop because of its high density of 350,000–700,000 plants per hectare and 100 flowers per plant. They studied 16 grasslands that were isolated by at least 230 m from oilseed rape fields, 17 grasslands and 17 oilseed rape fields that were 1–17 m of each other, and 17 oilseed rape fields that were isolated by at least 570 m from grasslands. The oilseed rape fields were all very similarly managed and the percentage of flower coverage did not differ significantly between isolated and adjacent grasslands, so those factors could not have affected the results. Fields with *Osmia bicornis* in traps were included for brood numbers and pollen analyses, but the isolated oilseed rape fields did not have significant numbers and thus were not analyzed. Using GIS software, the proportions of oilseed rape fields and of grasslands in landscape circles with radii of 250, 500, 700, and 1000 m were calculated.

The authors set up trap nests just inside the edges and in the center of the fields to test for brood numbers of *Osmia bicornis* and for the percentage of oilseed pollen in larval food. The nests were set up in March and all the nodes from the nests were collected and examined for brood cells after the end of oilseed rape flowering in May. Females of *Osmia bicornis* can establish 30 brood cells and forage for pollen up to 600 m away from the nests to provide for the larvae. Thus, high numbers of brood cells could be due to numerous females in the species or a high preference for nesting in that place. The authors conducted pollen analyses from 36 sites where *O. bicornis* was found, and these numbers were summed within the traps for each site and then the percentage of oilseed rape pollen per brood cell was averaged over all brood cells in a site to find the percentage of pollen in the larval food.

A generalized linear model with quasibinomial errors and the predictor presence of adjacent grassland was used to rule out the effect of grasslands on the presence of *O. bicornis* in oilseed rape fields. ANCOVA's were used to evaluate if the number of brood cells were higher in isolated or adjacent grasslands and oilseed rape fields. The authors used ANCOVA's to analyze the effect of local and landscape-scale availability of oilseed rape on the percentage of pollen in larval food. Linear regression models with the dependent variable as the number of brood cells were used to assess whether the number of brood cells increased with increased percentage of pollen in larval food. All models and tests were reflected in bar graphs and linear regression graphs.

The presence of grasslands had a positive effect on the number of brood cells in nests in 59% of adjacent oilseed rape fields. Only 12% of the isolated oilseed rape fields had brood cells. Additionally, the mean number of brood cells was 59% more in adjacent grasslands than isolated grasslands, and this was not affected by the presence of oilseed rape in the landscape scale. The percentage of pollen in larval food was higher in oilseed rape fields and adjacent grasslands than isolate grasslands and it did not differ between oilseed rape fields and adjacent grasslands. The number of brood cells increased with an increasing percentage of oilseed pollen in pollen food and was not affected by the availability of other food sources in the landscape-scale. These results suggest that mass-flowering crops positively affect the abundance of wild bees because they increase the access to food resources and nesting habitats. However, more research must be conducted because the increase in brood cells could be due to an increase in numbers of females of *O. bicornis*. The positive correlation may not reflect a positive effect on the biodiversity of the habitats around mass-flowering crops because *O. bicornis* is a generalist species, and thus may outcompete other species for nesting areas.

Plant-Pollinator Phenological Synchrony Stabilized by Biodiversity Against Climate Change

Biodiversity has been linked to the protection and sustenance of ecosystems against the loss of individual species. Studies have found that climate change, a contributor to the loss of species, has caused significant changes in phenology, mostly in species active in the spring. The biodiversity insurance hypothesis has never been expanded to include phenological synchrony as a possible buffer against the loss of individual species due to climate change. Bartomeus *et al.* (2013) investigated the phenological changes of wild bee species and of commercial apple crops over 46 years to find if bees and apples had phenological synchrony and if this was related to the richness of pollinator species. Using a contemporary data set, the authors picked pollinators that most frequently visited apple and tested for their phenological complimentarity. Bee and apple data were compared over time to find phenological mismatch and the rate of phenological change for different species with respect to apple bloom. Phenological synchrony was then tested against wild bee biodiversity. Phenological synchrony was found to increase with increasing biodiversity of the bee species and stabilize over time even though the rate of phenological shifting differed between species.

In order to test for the effects of climate change on the phenological synchrony of pollinators and apple, Bartomeus *et al.* compared independent data of commercial apple and wild bee species from New York, USA over 46 years. Apple was used because it blooms early in the spring and plants with early blooms have been found to have the most phenological shifts due to climate change's increasing temperatures. Twenty-six wild bee species were chosen to study by sampling the pollinators visiting commercial apple crops in New York between 2009 and 2011. Rare species and managed species were eliminated from the study to assure that all species are significant contributors and that phenological shifts were not due to managed reproduction. Geographical limits on the longitude and corrections on the effect of latitude were conducted prior to data analysis to account for geographical differences between apple orchards.

Statistical analysis was performed to find the phenological complimentarity of pollinators and plants from the present-day data. To measure the rate of phenological change for bee species and apple, the slope of the peak bloom was compared against the year. Phenological asynchrony was determined by computing the difference between the date of bee specimen collection and the date of peak apple bloom for that year. Any bees that could not have interacted with apple because their active time fell outside of the peak blooming period for apple were left out of the analysis. Additionally, the authors created a simulation analysis to investigate the effects of pollinator species richness on plant-pollinator phenological asynchrony and the stability of the asynchrony over time (46 years). The 26 bee species were randomly sampled to created communities with different species richness levels and for each richness level, a regression analysis of phenological asynchrony of the species against the year was conducted. Slopes closest to zero from the regression analysis indicated stability of pollinator and apple bloom asynchrony over time.

Bartomeus *et al.* discovered that there was phenological complimentarity between bee species from 2009 to 2011, so the data for different species could be compiled to compare phenological changes between wild bee species and apple. Both apple and wild bee species advanced the bloom peak and active periods, and mean April temperature also increased. There was no change in the degree of asynchrony over time, which suggests that the bees and apple had a stable level of phenological synchrony. While the mean phenological synchrony was very high (slopes were close to zero), different bee species had different phenological drifts over time.

Overall, however, the differential rates of phenological drifts evened out to stabilize the phenological synchrony. From the simulation of richness of bee species, the authors found that the higher the level of richness of bee species, the higher the baseline phenological synchrony and the more stable phenological asynchrony. These results indicate that even with the effects of climate change on the phenologies of apple and bee species, the plant-pollinator phenological synchrony for activity remains stabile. Additionally, increased biodiversity of pollinators has the effect of maintaining the pollinator-plant relationships by increasing phenological synchrony and stabilizing it overtime.

Decreasing Honey Bee Pollination of Watermelon Crops due to Climate Change buffered by Native Bee Pollination

As the Earth increases in temperature due to climate change, many temperature-sensitive pollinator species, such as honey bees, have declined in population and in pollination services. Since insect pollinators help pollinate 75% of the major global food crops, this decline in pollination services in recent years is especially concerning. Rader *et al.* (2013) studied the impact of climate warming on the pollination of watermelon crops, a major global crop, by different wild and manage honey bee taxa. Most studies correlating climate change with honey bee pollination have only investigated current trends and honey bees, and lack predictions for the effects of increased temperature on the diurnal activity patterns of specific bee pollinators. Using data on the visitation of watermelon crops by pollinator species, the different pollinator taxa present, and the amount of pollen on the stigma of watermelon flowers, the authors created models to predict how the pollination services of separate bee taxa would change. Varying possible climate change scenarios were accounted for using temperature records from weather stations near each of the 18 watermelon farms. Under the most extreme climate change scenario, the managed honey bees declined by 14.5% by 2099, while the wild bees increased so that the overall pollination services increased.

Rader *et al.* chose to collect data specifically from watermelon crops because they are completely dependent on insect pollination for fertilization and since they only bloom for one day. In addition, watermelon flowers attract a very diverse set of pollinators, which the authors accounted for by netting from the same sites on the same days as counting visitation rates and counting three species by wing during data collection. Eight dominant species groups accounted for 98% of all pollinators' visits. The group 'small dark bees' was the most diverse in species, containing 14 species, whereas the other groups were dominated by one species.

Pollinator visitation rates were determined by counting visits to watermelon over 45 seconds at 40 groups of flowers along a 50-m transect of crop row within each of the 18 watermelon farms in central New Jersey and eastern Pennsylvania. Each transect was observed three times between 8:00 and 13:00 hours on one day in 2007/2008, two days in 2005, and 3 days in 2010 at each farm. All days were sunny with wind speed less than 4.6 m/s to control for possible weather effects. In addition, the authors calculated pollination services by measuring the number of watermelon pollen grains on stigmas during a pollinator visit. Female virgin watermelon flowers were collected and individually placed with foraging bees, after which the pollen was allowed to attach to the stigma and then the number of pollen grains on the stigma were counted.

In order to analyze the data for pollinator visitation rates, pollination services, and species in comparison to temperature over time, the authors created pollinator response surfaces, which compared the time of day to the temperature and the flower visitation rate pooled for each taxon. Mean and variances for flower visitation rates within each grid for each pollinator taxon were calculated and the response surfaces where simplified to make them less sensitive to particular times of the day. Daily and yearly variations in temperature were incorporated for the current IPCC-based temperature predictions, and for the extreme and conservative future temperature scenarios. The authors created a Monte Carlo model, combining the future temperature predictions, the number of visits by each taxon, and the amount of pollen deposited per visit as a function of time of day, to estimate the changes in pollination services for each pollinator taxon over time. Within-species variation was accounted for by estimating visitation rates of only taxa known to be present during each time interval and at each specific temperature.

Rader *et al.* found that there were no common trends in pollinator activity patterns or for pollen deposition for the pollinator taxa. The pollinator response surfaces showed that the taxa responded differently to both temperature and time of day and pollination efficiency differed among bee taxa. For the pollen services models showing pollen deposition and visitation rate against temperature and time, some taxa were predicted to increase in their pollination services, while others contributed less in the future. The range between pollination services was extreme, with *Ceratina* increasing by 86.4% and *A. mellifera,* the managed common European honey bee, decreasing by 14.5%. While the managed honey bees decreased in pollination services, the overall change in pollination services was predicted to be +4.6% by 2094–2099 under the extreme climate scenario. This increase was due to the native, wild bee taxa contributing more under higher temperatures. If the wild pollinators that contributed to the increase in pollination services were not present in the model, the pollination services were predicted to decline by 15.3%, highlighting the importance of pollinator biodiversity and wild pollinators for maintaining global crops like watermelon.

Conclusions

The current studies on bees and other plant pollinators reveal the need for a change in pesticide use because of their impairment of functions and causation of fatalities. In addition, the link between climate change and the negative impact on pollinators has been strengthened by recent studies that span over 100 years and investigate network structures. However, even with climate change, wild pollinators have maintained pollination of mass crops, and mass crops have increased brood numbers of bees because of the increase in food for the bees. Plus, with increased biodiversity of bee species, the negative effects of climate change on pollination services are balanced out. These studies provide an optimistic view of the future of agriculture, but give lots of room to improve pollination of crops, use or create less poisonous pesticides, and combat the effects of climate change.

References Cited

Bartomeus, I., Park, M., Gibbs, J., Danforth, B., Lakso, A., Winfree, R., 2013. Biodiversity ensures plant–pollinator phenological synchrony against climate change. Ecology Letters 16, 1331–1338.

Burkle, L.A., Marlin, J.C., Knight, T.M., 2013. Plant-pollinator interactions over 120 years: loss of species, co-occurrence, and function. Science 339, 1611–1615.

Garibaldi, L.A., Steffan-Dewenter, I. Winfree, R., Aizen, M.A., Bommarco, R., Cunningham, S.A., Kremen, C., Carvalheiro, L.G., Harder, L.D., Afik, O., 2013. Wild pollinators enhance fruit set of crops regardless of honey bee abundance. Science 339, 1608–1611.

Girolami, V., Marzaro, M., Vivan, L. Mazzon, L., Giorio, C., Marton, D., Tapparo, A., 2013. Aerial powdering of bees inside mobile cages and the extent of neonicotinoid cloud surrounding corn drillers. Journal of Applied Entomology 137, 35–44.

Holzshuh, A. Dormann, C.F., Tscharntke, T., Steffan-Dewenter, I., 2013. Mass-flowering crops enhance wild bee abundance. Oecologia, 172:2, 477–494.

Palmer, M.J., Moffat, C., Saranzewa, N., Harvey, J., Wright, G.A., Connolly, C.N., 2013. Cholinergic pesticides cause mushroom body neuronal inactivation in honeybees. Nature communications 4, 1634. <http://www.ncbi.nlm.nih.gov/pmc/articles/PMC3621900/>

Rader, R., Reilly, J., Bartomeus, I., Winfree, R., 2013. Native bees buffer the negative impact of climate warming on honey bee pollination of watermelon crops. Global Change Biology 19, 3103–3110.

Williamson, S.M., Moffat, C., Gomerall, M.A., Saranzewa, N., Connolly, C.N., Wright, G.A., 2013. Exposure to acetylcholinesterase inhibitors alters the physiology and motor function of honeybees. Frontiers in Physiology 4. <http://www.ncbi.nlm.nih.gov/pmc/articles/PMC3564010/>

14. Global Climate Change: Lost Species and Lost Distributions

Cameron Lukos

It is well understood that global climate change is having effects on various species around the world. But what is global climate change and how is this affecting species. Global climate change is the term used to describe long term changes in weather patterns, temperature, winds, humidity, etc. This includes the effect of global warming. It has been suggested to cause species range shifts and extinctions. A species range shift is the change in geographic distribution of populations of a given species, most likely due to some sort of environmental change (e.g., climate) that can no longer be tolerated by that species. Shifting ranges have far-reaching effects on ecosystems as well as the species itself. Some species may have mechanisms and tools that prevent them from going extinct while others will not be so lucky. In order for us, as humans, to understand how species will survive or not, we must study the mechanisms and traits that are at species' disposal as well as look at the methods we use to study these mechanisms and traits. The following eight paper synopses describe (1) studies that analyze species mechanisms and traits to cope with climate change, (2) ways scientists are studying these mechanisms, and (3) how we can improve on what has already been done to help focus on conserving and preserving habitats and species from going extinct.

Climate Events Synchronize the Dynamics of a Resident Vertebrate Community in the High Arctic

In studying climate, scientists have been furthering their understanding of how climate events have been affecting a particular species. But it is unclear how climate will affect communities of species as a whole. Using the high Arctic as a case study, Hansen *et al.* (2013) describe changes in the weather patterns of Svalbard and how these events synchronize population fluctuations across the entire vertebrate community and cause a lagged effect on a secondary consumer, the Arctic fox. The synchronization of high Arctic populations is theorized to occur when winter rains turn to ice, causing the vegetation to be encased in ice and therefore unavailable to herbivores. This indirect bottom up effect drives population dynamics across the four vertebrates in Svalbard. With global warming, the frequency of winter rains and the subsequent icing of vegetation is expected to increase in the high Arctic and therefore strongly affect terrestrial ecosystems. Hansen *et al.* used statistical data based on population fluctuations and weather to demonstrate the effects of severe weather events on mixed populations.

The community Hansen analyzed was the island of Svalbard. The ecological community that he studied consisted of three herbivore species: Svalbard reindeer, Svalbard ptarmagin, and vole, and the secondary consumer the Arctic fox. Hansen *et al.* found correlated population fluctuations of all four species. The Arctic fox data were advanced one year due to a delayed population reaction to the change in herbivore populations. Based on these data, Hansen *et al.* hypothesized that the climate events were affecting the plant species which limited the amount of forgeable food for herbivores. This creates a bottom up effect which then causes dips in the Arctic fox population. In order to test this hypothesis, Hansen *et al.* ran linear regressions that modeled population growth rates as a function of population size and precipitation events.

The authors determined that after factoring in density dependence, the number of rainy days during winter months was the best predictor of annual population growth rates across species. Winter rains caused a negative effect on all species. Hansen *et al.* also found that summer temperatures had a positive effect on species growth rates. This confirmed their hypothesis that climate events do enforce synchrony among herbivores and causes the lag response of Artic foxes. Increased summer temperatures increase green foliage that, in turn, supplies more food for herbivores and thus an increase in the number of secondary consumers. Winter rains reduce access to food supply for the herbivores, causing increased mortality rates among old, sick, and very young individuals. The changing mortality rates cause a decline in population followed by an increase due to less competition and in the case of ptarmagins and voles, a drop in predation by the reduced population of the Arctic fox.

Evolutionary Rescue from Extinction is Contingent on a Lower Rate of Environmental Change

Lindsey *et al.* (2013) used the bacteria *Escherichia coli* to prove that rapid climate change will hinder the process of natural selection and may therefore cause extinction. Although some experiments have shown that slower rates of environmental change have led to more adapted populations or fewer extinctions. Lindsey *et al.* used different concentrations of the antibiotic, rifampicin with *E. coli* over different time periods to simulate slow, intermediate and rapid environmental change. They then genetically modelled all possible combinations of mutations that can result from slow rates of environmental change. The assessment of the engineered strains show that certain genotypes are evolutionarily inaccessible under rapid environmental change, and that rapid change could eliminate entire sets of mutations as options. They further speculated that intermediate levels of change might enhance expressions of genes, then there could be more endpoints as a result. This could have an effect on rates of adaptation that could among other things, increase rate of development of antibiotic resistance.

The experiment was carried out by propagating 1,255 populations of *E. coli*. The experimenters also created mutants to add the secondary factor of selective accessibility. These populations were then placed under increasing amounts of the antibiotic rifampicin. The treatments had different rates of change, ranging from sudden to gradual. The populations all started in an antibiotic free environment and ended at a maximum concentration of 190µg/ml. The populations that were subject to rapid change were exposed to the maximum rifampicin concentration after the first transfer and remained at that level for the rest of the experiment. Populations exposed to moderate change received the maximum amount of rifampicin halfway through the experiment and the populations exposed to gradual changed experienced the full amount of rifampicin on the last transfer of the experiment.

The results indicate that as the rate of environmental change increased the number of populations that survived the whole experiment decreased. Populations exposed to gradual change were able to become resistant to the antibiotic because there was more time for mutations to occur and spread. They also found that there were significant differences in growth rates of the populations exposed to gradual and sudden change, indicating that different mutations occurred in different treatments. For instance, in the case of the sudden populations only one mutation was detected, while in the moderate and gradual populations many different mutations were discovered. Lindsey *et al.* found a clear historical contingency for mutations that occurred in the intermediate environments. It suggests that the mutations allow for the lineage to gain other mutations, a historical contingency. From this, the results suggest the high rates of climate change may cause problems for species resulting in higher extinction rates. Not only will it wipe out genetic diversity but it may also result in the loss of potential mutations that could occur under less extreme conditions.

Population Growth in a Wild Bird is Buffered Against Phenological Mismatch

Reed *et al.* (2013) conducted a study to test whether the population growth of *Parus major* (great tit) was negatively affected by climate change, specifically, to see if climate change induced a phenological mismatch. They took four decades of life history data from a population of great tits in the Netherlands, whose breeding has evolved to coincide with the development of caterpillar populations, a food source for the young tits. Warm weather has a positive effect on the development of caterpillars that, in turn, affects the breeding success of the great tit. Due to warmer springs there has been a mismatch of the breeding time and the food peak creating an intensification of directional selection to earlier laying dates. However, this mismatch has not affected the population growth. Reed *et al.* demonstrate a mechanism that contributes to the decoupling; that fitness losses due to the mismatch are countered by fitness gains due to less competition. The result implies that populations may be able to tolerate maladaptation from climate without immediately declining.

To conduct their study, Reed *et al.* studied a population of *Parus major* in relation to a food source of caterpillars. The focus area has experienced spring warming in recent decades due to climate change. The birds rely on caterpillars as a food source for their fledglings and so match their breeding patterns with the seasonal peak of caterpillars. The experimenters ran a statistical analysis to test how strong their connection is with the mismatch affecting population growth. When the results showed no statistical significance, the experimenters created a fitness variable to further understand the decoupling of population growth.

Their results show that warmer temperatures create a larger mismatch between caterpillars and the tits. But there was no statistical significance with either analysis. The mismatch had no statistically significant connection between the population growth and directional selection. The addition of fitness also did not yield any statistical significance. The lack of statistical significance reveals that there is something further that has not been accounted for. The question becomes why is population growth not lower in years a large fraction of females lay too late? Reed *et al.* discuss two reasons for this. One is that the food peak is much narrower than the distribution of breeding dates and so reproductive fitness cannot be high for all females every year. The experimenters show that in an early reproductive year relative to food peaks, females who start early produced fewer fledglings than females that started later. Those females that started in the middle had the highest reproductive output. In a relatively late reproductive year late females had lower reproductive output than intermediates. The ones who do the best in this situation are those that reproduce early. This pattern is most likely due to the fact that some pay a higher energetic costs when feeding their broods. The second reason is that fledgling production is counterbalanced by improved independent survival of offspring due to relaxed competition.

Ecological Niche Shifts of Understory Plants Along a Latitudinal Gradient of Temperate Forests in Northwestern Europe

In order for species to survive environmental change and avoid extinction, species have to be able to either track suitable environmental conditions or adapt to the changed environment. Whether and how species adapt to environmental change is largely unknown. Wasof *et al.* (2013) examined the realized niche width (ecological amplitude) and the realized niche position (ecological optimum) of 26 plant species to see how each changes in relation to one another. A realized niche is the actual space that a species inhabits and the resources it can access as a result of limiting biotic factors present in the habitat. The authors created the niche width from a beta diversity metric, which increases if the focus species co occurs with other species. Wasof *et al.* used a detrended correspondence analysis (DCA) to represent the locations of the niche positions and then developed their own approach to run species specific DCAs to allow the focal species to shift its realized niche while others stayed put. Wasof *et al.* concluded that none of the 26 plant species maintained their realized niche width and position along the latitudinal gradient. A few species shifted their realized niche width but all of the species shifted their position. Most of the species that shifted their position shifted their realized niche for areas where soil nutrients and pH were poorer and more acidic. The results suggest that these plants are locally adapting or have plasticity. The pattern casts doubt on the idea that realized niches are stable in space and time.

Wasof *et al.* created identical sampling designs across seven regions along an 1800 km latitudinal gradient from northern France to central Sweden and Estonia via Belgium, western and eastern Germany, and south Sweden. The gradient encompassed most of the temperate deciduous forest in northwest Europe. In each region they sampled five kilometer by five kilometer landscapes containing a set of deciduous forest patches. The landscapes all shared similar features in terms of patch morphology and relative proportions of meadows to agricultural land and forests. To assess the shifts in realized niches of species along the gradient Wasof *et al.* focused on a subset of understory plants species that were part of the species pool of all the regions. Twenty six species were sufficiently frequent to be used for analysis. The assessment used Ellenberg indicator values (EIV) to rank the plant species along their optimum for light, soil nutrients, soil pH, soil moisture, temperature, and continentality. The authors first calculated the mean EIV to estimate the environmental conditions. Then they computed the median and range EIVs and then used GLMs.

Wasof *et al.* found that soil nutrient and pH values decreased with increasing latitude but latitudinal changes were significant for soil moisture. They did not find any significant variation in environmental heterogeneity across landscapes for light soil nutrients and soil moisture but as latitude increased environmental heterogeneity in soil pH increased, but only marginally. The latitudinal changes in the niche width and position differed between species. The majority conserved a similar niche width along the latitudinal gradient; some showed a positive linear, concave or convex relationship with latitude. The DCA scores showed that 22 of the 26 species moved northward and their position shifted to nutrient-poorer and more acidic soils. Those that had a convex relationship showed the same relationship with nutrients and pH. The concave relationship was a split result with some showing a positive correlation with nutrients and acidity, and others, a negative relationship.

High and Distinct Range-Edge Genetic Diversity despite Local Bottlenecks

The genetic consequences of being at the edge of species ranges has been the subject of much debate. Populations that occur at low latitude ranges are expected to retain high unique genetic diversity. Less favorable environments that limit population size at the range edges may have caused genetic erosion that has a stronger effect than past events. This study by Assis *et al.* (2013) provided a test of whether the population declines at the peripheral range might be shown in decreasing diversity and increasing population isolation and differentiation. The authors compared population genetic differentiation and diversity with trends in abundance along a latitudinal gradient to the furthest extents of the range of a sea kelp, *Saccorhiza polyschides*. Assis *et al.* also looked at recent bottleneck events to determine whether the recent recoded distributional shifts had a negative impact on the population size. They found that there was decreasing population density and increasing spatial fragmentation and local extinction at the southern edge. The genetic data revealed two distinct groups and a central mixed group. As the authors had predicted there was higher differentiation and evidence of bottleneck at the southern edge but instead of a decrease there was an increase in genetic diversity suggesting that extinction and recolonization had not reduced diversity and that this may be evidence of a process of shifting genetic baselines.

Assis *et al.* focused on the Portuguese coastline. They divided the coastline into twenty five twenty five kilometer x twenty five kiolmeter cells. Their sampling was intensified at the three northern-most and southern-most cells, by dividing the cells into five kilometer cells. They took their samples at comparable depths during the summers of 2008 and 2010. To evaluate the variability in the distribution and abundance of *S. polyschides*, Assis *et al.* plotted the presence and absence with a list of historical geo-referenced occurrences. To test genetics, they used genomic DNA using a CTab method and filter plates. The allele sizes were scored using STRand software and put into classes using MsatAllele. Genetic diversity was determined using FSTAT and the data were placed in a linear regression model. Bottleneck effects were tested using two methods: 1. Heterozygosity excess 2. M-ratios.

Assis *et al.* showed that there is a persistence of high unique genetic diversity at the species range edge. They found a decrease in the density and an increase in fragmentation with latitude, but the hypothesis of a decrease in genetic diversity with decreasing density was not confirmed. Allelic richness and heterozygosity increased towards the more sparsely populated south edge and the southern sites were strongly genetically different. The results raise the question of why genetic population diversity was higher at a low latitude edge. One hypothesis was the occurrence of microscopic development stages, but this is not supported by any of their data. An alternative hypothesis is the persistence of suitable habitat refugia at the southern locations.

Effects of Local Adaptation and Interspecific Competition on Species' Response to Climate Change

Adaptations, and how effective the adaptations are, allow species to have varying geographic ranges. For instance, species that have the tolerance to survive in cold climates will be able to live and survive in those conditions while others who do not have that ability will not be found. This gives different geographic ranges for all species. But our world is experiencing global climate change which means that the environments species are presented with will also change. Bocedi *et al.* (2013) used models to include the effect of climate change coupled with species interactions to understand these changing dynamics. To do this, they created simulations of two competing species across a linear climatic gradient that changes at different latitudes and gets warmer. Bocedi *et al.* gaged reproductive success by the individual's adaptation to local climate and its location relative to global constraints. In conducting their experiment they found that in changing the strength of adaptation and competition, competition reduces genetic diversity and slows the rate of range change. They also found that one species can drive the other to extinction long after climate change has occurred. Weak selection of adaptation and low dispersal ability also caused a loss of warmer-adapted phenotypes and that geographic ranges became disjointed and lost centrally adapted genotypes.

In creating these models Bocedi *et al.* had to consider (1) environment and climatic tolerance, (2) population dynamics and completion, and (3) dispersal. To understand environment and climatic tolerances the experimenters ran simulations with an individual-based map lattice model. The grid was 200 rows by 200 columns with only 20% of the cells as suitable habitat. The created environment was given a linear climate gradient that increased by 0.075ºC/column. The gradient was then shifted up to simulate climate change. The two species were modeled with potential for local adaptation affecting their chance to reproduce at a given location. For the population dynamics and competition they added in the two species having discrete generations which reproduce based on an individual-based formulation of the Ricker model. The species were not competing for space, but depending on the strength of competition, species can reduce each-others' growth rate. Dispersal was kept at a constant equal to 0.2 and if the disperser arrived in a cell where habitat was unsuitable it died. The simulations were run 5 times using different parameter sets. The climate was held stable for 100 generations and then changed one row every two generations and then held stable again for another 100 generations. All populations were at carrying capacity in all suitable cells; each simulation had a parameter changed, such as the strength of competition.

The models were a way to illustrate the variety of ways for local adaptation and species interaction to mold and modify the response of species to climate change. The models showed a number of nonintuitive outcomes with the interplay of strength of interaction and strength of selection. Bocedi *et al.* observed three alternative responses of species to climate change. First, the interacting effect of climate change and a competition can drive locally-adapted species to extinction by eliminating the range entirely. Second, a competing species may continue through an episode of climate change by shifting its range, sometimes at a highly reduced distribution after the range shift. Third, a species can persist through climate change and high competition but will suffer large reductions of genetic diversity and thus allow only a warm or cool genotype to dominate the species as a whole. These results suggest that when considering species ranges in regards to climate, scientists should consider more heavily local adaptation. Species may lose the central portion of their thermal range, and climate change, local adaptation, and competition will reduce the genetic diversity of the species.

Predicting Range Shifts Under Global Change: The Balance Between Local Adaptation and Dispersal

Global climate change is causing long lasting effects on all of Earth's natural systems. A consequence of these changes is species range shifting. Accurately predicting these shifts is very difficult and many methods have been criticized. The standard bioclimate envelope models (BEMS) have been criticized as too simple because they do not incorporate biotic interactions or evolutionary adaptation. BEMs are widely used though. Kubish *et al.* (2013) wanted to determine the evolutionary conditions of dispersal, because local adaptation or interspecific competition may be of minor importance for predicting future shifts. They used individual-based simulations at two different temperatures as well as competing simulations. Their results show that in single-species scenarios excluding adaptation, species follow optimal habitat conditions or go extinct if their connection to the environment becomes too weak. With competitors, their results were dependent on habitat fragmentation. If a species was highly connected to its habitat, the range shifted as predicted; if a species was only moderately connected to its environment, there was a lag time, and with low connectivity to the environment, the result was extinction.

Based on this work, Kubisch *et al.* determined that the BEMs may work well as long as habitats are well connected and there is no difficulty dispersing.

Kubisch *et al.* used simulations to simulate one species and two species interactions. The simulations are initialized with spatial separation of species. With two species scenarios, the colder half of the world is solely cold tolerant species and the warm half is solely warm tolerant species. In the single species scenarios they used only the warm adapted species restricted to their half. The experiments covered 3000 generations and each 1000 generations the temperature was increased. To test the influence of habitat connectivity the authors varied the dispersal mortality. For example, having high dispersal mortality means that the species has low habitat connectivity. The authors analyzed the data by calculating the range border position for the species and the number of occupied patches.

The authors determined that with their single species simulations that with low dispersal mortality the species will initially increase its range slightly and then follow the predicted shift. If the value for dispersal mortality is greater than 0.7 then the population completely collapses. The position of the range border is determined then by the niche width and the gradient. The authors then factored in mutations that would help adapt them to their temperatures and the resulting scenarios showed that with high connectivity the species takes over the whole range. With intermediate connectivity there was still an increase in expansion but not as quickly. The wide niches increased the survival of the populations even with a giant collapse at the same period of time. With two species systems the authors did not find any change in the speed of range shifts. By increasing the dispersal costs with this as well they saw an increased lag in the shift border.

Limited Evolutionary Rescue of Locally Adapted Populations Facing Climate Change

The role of dispersal is key factor in a population's evolutionary potential. Dispersal is how species are able to spread out and occupy different ranges as well as add potential for new species to evolve. Dispersal facilitates the proliferation of beneficial alleles throughout the range of species populations and increases adaptation. But, when habitats are heterogeneous and individuals become locally adapted, dispersal may reduce fitness by increasing poor adaptation. An experiment conducted by Schiffers *et al.* (2013) used an allelic simulation model to quantify the effects of dispersal on a population's evolutionary response to climate change. Their results showed that the relationship between gene flow and heterogeneity may decrease effective dispersal, and result in a population size that substantially reduces the likelihood of evolutionary rescue. Even when evolutionary rescue occurs, the species range after climate change may be narrowed and so the rescue is only a partial fix. This implies that local adaptation without consideration of non-climatic factors may be an overestimation of the population's evolvability.

To test their hypothesis, Schiffers *et al.* developed an allelic spatially explicit and individual based simulation model to test the interactive effects of gene flow and adaptation on populations experiencing environmental change. The modeled organism was a bisexual annual plant species with a cross-pollinating breeding system. The range size was thirty-two by thirty-two grid cells and the edges of both axes are joined together. The grid cells were described with two environmental conditions: local environment conditions are stable over time and climatic conditions like maximum annual temperature change over the simulated period. Climate conditions were homogenous across the space. The grid cells each could support a set carrying capacity that was constant across the region. All individuals were diploid and were located in continuous space. At each generation, reproduction with mutation, recombination, gamete dispersal, parental death, offspring dispersal, selection on survivability of juveniles, and density dependent mortality were all simulated. The simulated plant's genetic structure was composed of 15 loci for each of the two considered traits. The authors ran two different scenarios; one in which the loci were all situated on one chromosome, and one in which all loci were split. They also made sure that all individuals could potentially bear offspring. The final simulations were run with a constant temperature for the first 200 generations and then were increased by 2°C over the next 100 generations. After the period of change the authors assumed the climatic conditions to be stable at the end of 500 simulation years.

The results showed that under no environmental change, population size, individual fitness and genetic variance were stable over time except when dispersal distances were too small. With environmental change, the population size started to decline when the average individual phenotype lagged behind the optimum. When the mutation rate was set to zero the populations eventually died out. Populations' responses to rapid climate change fitted three general classes: complete evolutionary rescue, partial evolutionary rescue, and extinction. Complete rescue occurred when there were enough beneficial mutations and these mutations were able to spread unhindered across the landscape. Partial rescue occurred when there were beneficial mutations but the individuals were unable to fully spread across the landscape. Extinction occurred when there was a failure to have enough beneficial mutations. The effect of dispersal in regards to habitat heterogeneity was that dispersal generally had a negative effect on individual levels of adaptation. But the model results also confirmed the benefit of dispersal on a population's ability to adapt. Schiffers *et al.* showed that the evolutionary potential of populations in deteriorating conditions might be overestimated when neglecting the effects of local adaptation to heterogeneous habitats. This may be important due to increased habitat deterioration leading to reduced habitat availability, increased habitat fragmentation and stronger habitat heterogeneity which will impede the ability of species to track their preferred climate.

Conclusions

Hopefully after reading these synopses, you are better able to understand the importance of global climate change and how this is affecting species everywhere. You can see that species do have natural defenses that could save them from stressful environments but it may not be enough. There are many other factors that have to be considered and it is all not just doom and gloom for all species. In order for us to understand how climate change is affecting species modeling, better tools must be developed. Without understanding how climate change is affecting species how can we as humans expect to do anything to save these at risk species as well as ourselves?

References Cited

Assis, J., 2013. High and Distinct Range-Edge Genetic Diversity despite Local Bottlenecks. *PloS one* 8, 1932-6203.

Bocedi, G., Atkins, K., Liao, J., Henry, R., Travis, J., Hellmann, J., 2013. Effects of local adaptation and interspecific competition on species' responses to climate change. Annals of the New York Academy of Sciences 1297, 1749-6632.

Hansen, B.B., Grøtan, V., Aanes, R., Sæther, B.-E., Stien, A., Fuglei, E., Ims, R.A., Yoccoz, N.G., Pedersen, Å.Ø., 2013. Climate events synchronize the dynamics of a resident vertebrate community in the high arctic. Science 339, 313—315.

Kubisch, A., Degen, T., Hovestadt, T. and Poethke, H. J. 2013. Predicting range shifts under global change: the balance between local adaptation and dispersal. Ecography, 36 873–882.

Lindsey, H.A., Gallie, J., Taylor, S., Kerr, B., 2013. Evolutionary rescue from extinction is contingent on a lower rate of environmental change. Nature 494, 463–467.

Reed, T.E., Grøtan, V., Jenouvrier, S., Sæther, B.-E., Visser, M.E., 2013. Population Growth in a Wild Bird Is Buffered Against Phenological Mismatch. Science 340, 488-491.

Schiffers, K., Bourne, E. C., Lavergne, S., Thuiller, W., & Travis, J. M. 2013. Limited evolutionary rescue of locally adapted populations facing climate change. Philosophical Transactions of the Royal Society B: Biological Sciences, *368,* 1610.

Wasof, Safaa 2013. Ecological niche shifts of understory plants along a latitudinal gradient of temperate forests in north-western Europe Species' realized-niche shifts across latitude. Global ecology and biogeography 22 1466-822.

15. Ecological Restoration

Andrew Walnum

The idea of restoration has changed dramatically. During the late 19th and early 20th century restoration was being used to describe silviculture and erosion control not reestablishing past ecology and hydrology (Throop 2000). Although the idea of ecological restoration (sometimes referred to ecosystem restoration or restoration ecology) did not fully develop until the 1980s, it is quickly being recognized as important practice for combating local and global habitat destruction. There are several reasons as to why we should focus on restoring disturbed and degraded environments. Clewell and Clewell (2006) recognize five rationales for restoration ecology: technocratic, biotic, heuristic, idealistic, and pragmatic. Ecological restoration, when done correctly, can significantly benefit not only increased biodiversity but provide ecosystem services for humans. One such example is wetland restoration. It has been shown that restoration of wetland can significantly improve species richness, although the species structures (particularly for macroinvertebrates) can also change (Ilmonen *et al.* 2013). At the same time restored wetland can have positive economic benefits. Barbier *et al.* 2013) looked at the economic benefits of restoring wetlands southeast of New Orleans. They found that increasing the wetland cover and vegetation could save $99-$133 per unit area in destruction from storm surges, resulting in saving up to $592,000 to $792,100 in property damage per parish. The ecological and economic benefits from ecosystem restoration are numerous.

Understanding what ecological restoration entails and its goals are important not only for the public but also for practioners. Although every restoration project has the end goal of restoring an ecosystem to pre-disturbed state, how to obtain and measure successful restoration continues to become more refined. For instance, recognizing that ecosystems are not static but dynamic systems containing species with both wide and endemic ranges is essential for restoring habitat. The landscape in which an ecosystem resides must be taken into context in order to restore the land in a way that allows beneficial and native species to permeate and at the same time exclude exotic or invasive species (Shackelford *et al.* 2013). With global warming being accepted as an all too real and imminent threat, restoration ecologists must also find ways to allow ecosystems to change to different moisture and temperature regimes while also protecting current species.

Time and research are needed in order to find ways to properly restore ecosystems and restoration ecologists continue to gain information about proper restoration techniques and practices. Although ecosystems can be placed into specific categories it is important to study each individual ecosystem and understand what each one needs to be successful. References ecosystems are supposedly healthy habitats that are used as a baseline to compare the health of an ecosystem that is being restored. Recognizing that ecosystems still maintain differences based on location and that all ecosystems have been influenced by humans to some degree must be taken into account (Halme *et al.* 2013). Understanding the kind of microclimates certain species require for survival is time intensive but important for understand proper vegetative structure. Despite the large amount of knowledge we still require many successful projects have been implemented. Steen *et al.* (2013) found that implementing past burning regimes in Longleaf Pine sandhills helped to increase the population of the lizard *Aspidoscelis sexlineata*, an important indicator species for the ecosystem.

Ecological restoration has been viewed has a global imperative by the United Nations Convention on Biodiversity for achieving biodiversity goals from the Aichi Biodiversity Targets conference for 2020 (Aronson and Alexander 2013). It is also recognized as necessary for helping to mitigate food and water shortages along with increasing the livelihood of the global population. The CBD calls for restoration practitioners from around the world to come together and share knowledge needed to combat global issues on habitat destruction. It also calls for nations to implement policies that are necessary for environment sustainability.

Ecological restoration will continue to grow into a more refined practice. By better understanding species composition, vegetative structure, and other ecosystem functions we will be able to more successfully restore and protect degraded ecosystems. The socioeconomical and ecological benefits we stand to gain in communities, local and global, are too important to ignore as we head to an unpredictable future.

Ecological Restoration for the 21st Century

Ecological restoration is a relatively new and evolving field that developed in the 1980s and focuses on restoring land that has been degraded or destroyed by human activities. This study and practice is gaining popularity by governments, private businesses, and community organizations that recognize the importance of having healthy ecosystems for intrinsic, practical, and economic benefits. However, in order for any ecological restoration project to work there must be a firm set of goals in place so that practitioners can plan and act accordingly to achieve ecosystem health. The Society for Ecological Restoration International Primer on Ecological Restoration (Primer for short) defined and detailed these goals in a section called “The Nine Attributes of Ecological Restoration” which places these attributes into four separate groups. Shackelford et al. (2013) attempts to build upon these goals as well as add a new group “The Human Element.” With a large increase in the amount of information about rebuilding ecosystems and restoring their health, changes to the nine attributes are needed for repairing environmental damage done to habitats by humans.

Shackelford and her colleagues found new ways to define ecological restoration goals and important categories by consulting with professors, practitioners, students, and post-doctorates versed in the study. These volunteers were placed into small individual groups for a literature review and discussion on articles pertaining to restoration ecology. Then, a larger group discussion was held with the participants to talk about key points that arose from reviewing the literature. The main points of the discussion groups were recorded in order to find what the experts or students felt were important to the field of ecological restoration.

The results from the discussions raise new thoughts on what the nine attributes of ecological restoration should be. Before, the list was as follows: 1. The restored ecosystem contains a characteristic assemblage of the species that occur in the reference ecosystem and that provide appropriate community structure. 2. The restored ecosystem consists of indigenous species to the greatest practicable extent. In restored cultural ecosystems, allowances can be made for exotic domesticated species and for noninvasive ruderal and segetal species that presumably co-evolved with them. Ruderals are plants that colonize disturbed sites, whereas segetals typically grow intermixed with crop species. 3. All functional groups necessary for the continued development and/or stability of the restored ecosystem are represented or, if they are not, the missing groups have the potential to colonize by natural means. 4. The physical environment of the restored ecosystem is capable of sustaining reproducing populations of the species necessary for its continued stability or development along the desired trajectory. 5. The restored ecosystem apparently functions normally for its ecological stage of development, and signs of dysfunction are absent. 6. The restored ecosystem is suitably integrated into a larger ecological matrix or landscape, with which it interacts through abiotic and biotic flows and exchanges. 7. Potential threats to the health and integrity of the restored ecosystem from the surrounding landscape have been eliminated or reduced as much as possible. 8. The restored ecosystem is sufficiently resilient to endure the normal periodic stress events in the local environment that serve to maintain the integrity of the ecosystem. 9. The restored ecosystem is self-sustaining to the same degree as its reference ecosystem, and has the potential to persist indefinitely under existing environmental conditions. Nevertheless, aspects of its biodiversity, structure, and functioning may change as part of normal ecosystem development, and may fluctuate in response to normal periodic stress and occasional disturbance events of greater consequence. As in any intact ecosystem, the species composition and other attributes of a restored ecosystem may evolve as environmental conditions change.

All of these attributes are meant to describe what a practitioner of restoration ecology should strive for when restoring an ecosystem. These nine attributes fit into the categories of species composition, ecosystem function, landscape context, or ecosystem sustainability. The authors also talk about the importance of fifth category, the human element. Humans play an integral part in shaping ecosystems and their involvement should be taken into account during restoration. The paper argues that in some cases restoration should be set to include permanent human involvement as can be seen in places like Europe where grasslands have been maintained for centuries by grazing animals or mowing. Social and cultural values must also be taken into account, especially in urban areas or areas that are considered important or sacred to a group of people.

The authors go on to add improvements to the original four categories. For species composition, they suggest recognizing that animals and plants are dynamic and that "indigenous" or "native" species can have large ranges. Also, using historical references for species composition may not be practical but rather looking at current, similar ecosystems to gain a better perspective about what species need to be a part of the restored habitat. When looking at ecosystem function, the authors argue that rapid climate change must be taken into account as the historical ecosystem did not have to face to same stresses or benefits that global climate change might raise. In addition, the economic or social services that a restored ecosystem can provide should also be taken into account. Using more trait-based measurements of ecosystem stability is also a new suggestion. They also explain the difference between resistance and resilience. Whereas resistance refers to an ecosystem being able to remain even through a large disturbance resilience is how an ecosystem is able to handle smaller disturbance over a long time period. For landscape context, it is important to understand the permeability of the landscape or how species and genes can spread through an environment. Knowing key areas that might allow invasive species, predators, or disease to enter can be taken into account and the ecosystem can be planned to minimize negative effects. Overall, the revision of ecological restoration goals can allow for further improvements to restoring an ecosystem that might not have been taken into consideration before.

Ecosystem Restoration is now a Global Priority

Conservation and sustainable use cannot guarantee stable ecosystems on a human-dominated planet. Restoration ecology can help us to obtain goals of a sustainable future especially as global knowledge of the practice continues to grow and become more refined. The United Nations Convention on Biological Diversity (CBD) held in Hyderabad, India in 2012 urged countries and relevant organizations and businesses to incorporate ecosystem restoration. Although many future goals may seem some lofty, they are realistically attainable with a proper adoption restoration at the legislative level (Aronson and Alexander 2013). Just as important, the need to be cautious about knowing current limits of restoration must be taken into account for large-scale goals.

Aronson and Alexander worked on highlighting the information from The United Nations Convention on Biological Diversity. They used relevant statements and themes from the past year and the convention in Hyderabad that focus on mainstreaming ecological restoration in governmental policy. They organized these highlights into four sections: "Ecosystem Restoration as a Conduit for Achieving Multiple Objectives," "Ecosystem Restoration Day at the Rio Convention Pavilion," "The Hyderabad Call and COP11 Decision on Ecosystem Restoration," "A Way Forward: Global Partnership for Local Results."

For the first section, "Ecosystem Restoration as a Conduit for Achieving Multiple Objectives," the authors state how restoration is now recognized as an essential for achieving goals set for by the CBD and Aichi Biodiversity Targets for 2020. Although it has been recognized by the CBD that restoration cannot replace conservation, ecosystem restoration is agreed to be necessary to attack "poverty alleviation, assuring food and water security, and generating sustainable livelihoods". As restoration ecology has progressed, restoration techniques have become globally recognized. This increase of knowledge has led to The International Union for the Conservation of Nature (IUCN) to release a new volume entitled: "Ecological restoration for protected area: principles, guidelines and best practices.

The "Ecosystem Restoration Day at the Rio Conventions Pavilion" first explains how each of the Conventions of the Parties (COP) hold their Conventions to find synergies and collaborative work programs. An entire day of the Hyderabad convention was devoted to laying the foundation for the Hyderabad Call which mainly focused on finding cost-effective restoration outcomes at all levels of restoration goals (local, national, and global). Fourteen partners supported the Hyderabad Call for a Concerted Effort on Ecosystem Restoration including the three countries that are hosting the Rio Conventions: India, South Korea, and South Africa. The Hyderabad Call recognizes that a combined global effort is needed to achieve the Aichi Targets for 2020 by implementing programs that also focus on the goals of other conventions including mitigating climate change, ecosystem-based adaptation, reducing land degradation, and proper wetland use that fit with global development goals. Making a "...concerted and coordinated long-term efforts to mobilize resources and facilitate the implementation of ecosystem restoration activities" is the main point of the Hyderabad Call. At the end of the CBD COP11, Decision XI/16 outlines how these goals can be obtained. These include: capacity building initiatives, which include regional workshops and technical training courses; knowledge sharing through searchable databases, including e-learning modules, case-studies, and best practices; exchange programs among agencies, restoration practitioners, and researchers; awareness-raising and communications outreach on the economic, ecological, and social benefits of ecosystem restoration including the general public, policymakers, and environmental managers; and integration of ecosystem restoration into broader planning processes.

"A Way Forward: Global Partnerships for Local Results" focuses on finding ways to bring the restoration community together and financing active restoration projects. The authors recognize that national commitments are the "enabling factor" for providing funding. A coordinated effort of government, NGOs, and corporations is needed from local to global levels to provide open access information on restoration, guidance for projects, the tools needed to perform restoration activities and appropriate technologies. Also, the Hyderabad Call states that "research should not only improve our ability to undertake restoration activities so as to protect biodiversity, but also contribute to capacity-building, sustainable livelihoods, and improving the human condition in general." An experienced global community of restoration ecologists is also an important part to obtaining sustainability goals.

To conclude, the authors stress that the community cannot make promises about restoration that they cannot keep. There are large obstacles that must be faced such as implementation of and financing that must be overcome to achieve large scale restoration. However, there is a great potential to achieve global goals in ecosystem restoration as both bottom-up and top-down projects show hope for attracting future government and private investors.

Challenges of Ecological Restoration: Lessons from Forests in Northern Europe

The degradation of ecosystems around the world continues to occur and an increasingly rapid rate. As a relatively new field of ecology, ecological restoration sometimes struggles to find ways to combat the challenges faced by restoring disturbed ecosystems on a local and global scale. At the latest Convention on Biological Diversity (CBD, 2010 in Nagoya, Japan), restoring ecosystems was recognized as one of the most important tools for preventing future loss of bio diversity. Although this field has grown rapidly since the 1980s, many developed countries, especially in Northern Europe, are just starting recognize its importance. Several challenges are needed to be overcome in order to protect biodiversity on a large scale. This study uses northern forests in Europe as an example on what can and needs to be done in order to ensure long-term environmental and biodiversity preservation to reach goals set forth by the CBD.

The basis of this paper comes from discussions held in workshop hosted by PRIFOR; a Nordic work group that focuses on primeval boreal forests. The authors stress that although these discussions concentrated on facing restoration challenges in northern European forests, that the conclusions reached during the workshop can be viewed as a general reference for restoring various forests. They discuss "...the objectives, theory, practice and problems related to ecological restoration of forests." After summarizing all of these aspects they next identify future challenges that will be faced in forest restoration.

Anthropogenic activities have significantly affected the structure and composition of forests, especially in areas that have been inhabited for long periods of time. Much of this is due to logging or clearing land for agriculture. The overall complexity of forest structure has seen a huge decline resulting in large amounts of habitat loss. For instance, the accumulation of deadwood in old-growth forests led to increased habitat space for fungus and arthropods. These changes have created new species assemblages resulting in more generalist species. It is important to recognize that over that all forests have been shaped by human intervention and are at varying states of "naturalness" and restoration plans must react accordingly.

Currently there are a large number of restoration projects taking place in northern Europe. These projects focus on restoring or improving past ecological function and structure of the forest. Many of these projects focus on restoring the hydrology of an area by blocking off drains using alluvium to raise the water table. Creating open canopies allow for certain trees and plants to reestablish within the forest. Another important implementation is returning regular burning regimes to the ecosystems as was seen in the past. These processes are expensive and represent an extremely small fraction of the total northern forests. However, they help to restore biological diversity to areas and assist in allowing the forest to behave naturally.

An important aspect of restoration ecology is understanding what cannot be restored in a certain ecosystem or location. Climactic, ecological, economical, and sociological challenges may be present that prevent full restoration from taking place. For example, at the forest level, burning may help to reestablish certain plant species but it providing funding for consistent fire regimes is most likely economically out of reach. On the species level, there is a large amount of information that is needed in order to fully maximize the forests ability to maintain a high level of species richness. Practitioners must understand the types of habitats and microclimates needed to restore species richness along with finding corridors that allow species to naturally reestablish a forest ecosystem. In addition, restoration may be entirely feasible in a particular location but is not allowed to occur due to unwanted side-effects such as increased pest numbers and erosion. Currently, it hard to measure the effects of restoration on reaching long-term biodiversity goals. Most studies only show short-term trends in changes in species richness and focus on immediate small scale restoration goals. However, much evidence does support forest restoration as a viable option for increasing biodiversity.

Although still a young field, ecological restoration has learned many lessons as it strives to increase biodiversity and/or restore past ecosystems. For one, ecological knowledge should be focused on the particular target ecosystem that is being restored. Using generalized techniques may not reflect the natural disturbance regimes of the past resulting in little to know changes in biodiversity. Knowledge on each specific ecosystem must be gathered to ensure a complete understanding of restoration goals and practices. Just as important is the issue of defining "naturalness." Often times species are reintroduced into a target ecosystem based on a historical reference, such as an undisturbed forest. It is important to recognize that all ecosystems have been affected by human contact to various degrees and a "natural" forest may not fully reflect past species trends. Consistently monitoring an ecosystem is also very important for restoration purposes. Monitoring can help a practitioner decided how much effort must be needed to restore and ecosystem. Sometimes, an ecosystem can bounce back from a disturbance naturally and may not need a large amount of funding or resources to restore. Monitoring after restoration is also important for understanding species dynamics and increasing our knowledge of species function within an ecosystem.

There are several challenges relating to forest restoration that must be addressed in the future. The first being unpredictability. Restored ecosystems are often small fragments of a large ecosystem and may be prone to unpredictable stresses that may occur. The role of global climate change may have an unknown effect on a restored ecosystem. Another challenge is the time scale needed for an ecosystem to become self-functioning. These forests may take hundreds of years to reach a point where they no longer need human intervention in order to thrive with high levels of species richness. Despite current restoration efforts, we still do not know if restoration helps lead to a more resilient ecosystem. Social restrictions are also a large challenge that must be faced. It may be difficult to find areas large enough to support a sustainable, self-functioning ecosystem in many areas from fragmentation due to urbanization or landowners refusing to allow restoration on their property.

Despite these challenges, there are global initiatives being implemented to restore large areas of land to increase biodiversity. Restoration ecologists understand what is needed to properly restore ecosystems and there is increasing information about habitats and species assemblage. Planning across administrative levels will help ensure funding and implementation of restoration projects world-wide.

Detecting Restoration Impacts in Inter-Connected Habitats: Spring Invertebrate Communities in a Restored Wetland

The goal of every restoration project is to restore degraded ecosystems as closely as possible to their pre-disturbed functions. For wetlands, restoring the hydrological function of the area is usually what restoration ecologists aim to achieve, often at a rate which quickly makes changes to the hydrology and chemistry of the landscape. Although ecological restoration is an important growing field, very little is known about the inter-habitat effects of restoration. Freshwater springs regularly form along wetland ecosystems but there have been no studies to find how restoration might affect these habitats. Illmonen *et al.*(2013) looks at the effects of restoring wetland on these non-target ecosystems by looking at macroinvertebrate diversity. Because these habitats are geographically scattered the authors believed that recovery time for these springs may be slow due to poor dispersing mechanisms for macroinvertebrates, although more cosmopolitan species may take over quickly.

Ilmonen and colleagues from the University of Oulu, Ruuhikoskenkatu, and the Finnish Environment Institute used impacted and control sites along with independent sites away from the restoration sites to perform a before-after-control-impact (BACI) study. Ditches were filled to prevent draining and restore the ecosystem back to its pre-disturbed mire state. Control springs were 5-10 meters away from restoration attempts. Independent control springs were located 300 km southwest of restoration activities but within the same ecoregion. Samples were taken in May 2001 before restoration occurred. Subsequent samples from the springs were collected in May 2003, 2005, and 2010. The macroinvertebrates were sorted to their lowest taxonomic level and then sorted into two categories: "freshwater generalist" and "spring specialist." Using generalized linear mixed modeling, response variables of proportional abundance of crenophilous (spring and cold specializing) taxa compared to benthic invertebrates and total taxon richness which was rarefied to 100 indviduals.

Ensuing testing of macroinvetebrate richness after restoration indicated a large decrease in the number of crenophiles in both the impacted and controlled restoration sites. The marked decrease was more prominent in impacted sites then the controlled sites. However, by 2010, both sites had recovered their crenophile diversity to almost pre-restoration activity levels.

Ilmonen and colleaugues recorded an increase in the spring of a few generalist species, particularly *Psectrotanypus varius* and *Nemoura cinerea*. Following restoration, *P. varius* abundance increased from no individuals to an average of 107 individuals per spring in 2003. The average number of individuals then decreased sharply to mean of 8.5 and 2.7 in 2005 and 2010 respectively. The number of individual *N. cinerea* also increased immediately following restoration and again corresponded to a decrease in crenophiles. Using the rarefied species richness results, there was an overall negative trend for restored springs. Remote springs showed an increase in species richness over time.

Overall, there was little change in the macroinvertebrate community of remote springs whereas all springs in the restored area showed a significant change after restoration and slowly recovered to pre-restored levels. Changes in the community of restored springs correlated with increasing changes of moss cover and water depth.

The results show that restoration can have a harmful impact on not only restored springs but also neighboring springs being used as controls. Although there was little impact on species richness over time, the community structure of the springs was greatly impacted. The after effects of restoration include an increase in the number of generalist freshwater taxa and a less marked decrease in crenophilous taxa. The lentic (still water) loving *P. varius* quickly established itself in the restored and control springs shortly after restoration occurred. In addition, there was a change in the abundance of species and taxa indicating that measuring species richness does not accurately convey changes in the ecosystem post-restoration. Species richness may reflect little change in the community over time but does not account for species turnover and a restructuring of the macroinvertebrate community. A comprehensive analysis of species abundance is a more accurate indicator of ecosystem changes. Off-site control springs showed almost no changes in species richness or abundance indicating that measured changes were caused by restoration activities.

The results also indicate the importance of time scale for these restoration based studies. Changes in disturbances can vary over time along with a slow response by flora and fauna to restoration. A short sampling timescale may not allow enough time to measure biological and chemical changes in a restored ecosystem. Also, it is important to have non-impacted control sites to measure changes in restored and adjoining ecosystems against. For all future restoration projects a pre-restored species list should be conducted in not just the target ecosystem but neighboring ecosystems. Based on the results a restoration project may not be considered justified if there are any red-listed species that might be affected by it.

The Value of Wetlands in Protecting Southeast Louisiana from Hurricane Storm Surges

Wetlands are recognized as important habitats not only for their benefits of maintaining biodiversity, water purification, erosion control, and carbon sequestration, but also their ability to reduce the impacts of storm surges. Hurricanes pose a particular threat coastal areas as can be seen during Katrina and other devastating hurricanes. Wetland restoration in areas along the Gulf Coast seems to be a logical way to help reduce the devastating impacts of surges and floods from ocean storms. However, there has never been a full analysis combining the hydrological and economic impacts of increasing wetland areas along the Gulf Coast. The authors of this study used models to look at the effects of increasing wetlands on property damage in Southeast Louisiana, near New Orleans. Their study finds that an increase in 10% vegetation cover per square meter saves $99-$133 in property damage per unit area and only a 1% increase saves $24-$43.

Barbier and colleagues used storm simulations across a transect along with estimates of analysis of the economic impact of a storm surge. The transect was chosen using numerical models and the (ADCIRC) unstructured grid hydrodynamic model to predict the direction, intensity, and duration of storm surges. Twelve locations along the transect were collected using ADCIRC and were sub-sampled to create 100 points from sea to land. Next, the wetland-water ratio and bottom roughness along the transect was collected. The wetland-water ratio (W_L) was based on a scale of 0-1 with 0 being open water and 1 representing solid marsh. Bottom roughness (W_R) is the value of friction caused by vegetation with 0.002 being no vegetation and 0.045 being dense vegetation. Reducing surge power was then measured as the maximum amount of attenuation over each of the 11 transects between the 12 locations. Each one of the 11 transects was 6,000 meters long and the W_L and W_R along each transect was averaged. The authors were next able to change the W_L and W_R values to observe changes in storm surge attenuation. Changes in storm surge frequency and duration can vary greatly but the authors used expected damage function approach to find the marginal values of W_L and W_R on damage to surrounding human-inhabited areas.

The authors found a direct correlation between both increasing wetland-water continuity and vegetative roughness on storm surge attenuation. More and wetlands and vegetation decrease the intensity of incoming waves from storms. Increasing the wetland-water ration by only 1% reduced storm surge intensity by 8.4% to 11.2% and 1% increase in wetland roughness decreased storm surge by 15.4% to 28.1%. This reduction in storm surge also has an effect on the amount of money saved from damage reduction. A 10% increase in wetland-water continuity saves \$99-\$133 dollars per unit area and a 1% increase saves \$24-\$43 per unit area. If an increase in wetland continuity is expanded along the transect, the results are even more positive. An increase in wetland-water along the full length of a 6,000 meter transect results in saving \$592,000 to \$792,100 for the average sub-planning units in local parishes surrounding the wetlands. An increase in bottom roughness from vegetation accounts for \$141,000 to \$258,000 saved for the average sub-planning unit.

Although used for only one transect, the study helps illustrate the need for wetland protection in the future. Wetlands provide a large array of environmental services but there most important benefit may be the protection of coastal property. However, restoration is expensive and even with a large scale project along the Gulf Coast, there would continue to be a decrease in the number of wetlands over time. As more information on the economic benefits of maintaining wetlands comes out, it may prove to be more beneficial in the long-run to spend money on restoration to protect damage by storm surges.

Gallery Forest or Herbaceous Wetland? The Need for Multi-Target Perspectives in Riparian Restoration Planning

There has been a large preference for restoring forests as opposed to other restoration goals such as vegetation density and river processes. This penchant for trees can mostly be attributed to their aesthetic value and their cultural and biosociological benefits such as suitable places for restoration. In addition, the science used to judge riparian restoration often measure forest galleries as a sign of overall ecosystem health. This measurement might not take into account other factors that might prevent additional habitats such wetlands and ponds from being restored. Moving away from a single-minded focus to taking into account the landscape mosaic should be the new direction that restoration ecologists take for future projects. This new approach implies that forests are not planted in places along a river or stream system that they would be not be found in the past. Wetlands are often overlooked as important ecological communities in the context of riparian restoration and should receive equal attention during restoration efforts.

Weisberg and colleagues used case studies to determine the extent of wetland reduction in riparian forest restoration. They used studies and maps to look at this reduction over time. After sorting through the information they then proposed a new perspective for riparian restoration using previous studies. The information was combined to create an all-encompassing plan that can be used to restore not only riparian forests but the important wetlands that are often time ignored during restoration efforts.

The authors start by discussing the historical and ecological values of wetlands. In the past, wetlands were more widely spread across river systems. As much as 59%-98% of wetlands have been reduced along various riparian ecosystems in the U.S. In Switzerland, 90% of wetlands have been destroyed or converted to land for agriculture. Restored wetlands can have several benefits for riparian systems and adjacent agricultural land. Fluvial marshes are sometimes better at preventing bank erosion then riparian forests and provide important habitats for many different organisms. They also can trap stream sediments leading to reduced siltation in estuaries further downstream. Restored wetlands near land used for agriculture can benefit soil hydrology and hold water for extended periods of time, reducing the risk of downstream flooding.

In general, restoring wetlands is done at a local scale within an existing wetland community. Primarily, wetlands are restored by raising the local water table until ponding occurs. Water channels are plugged with alluvium preventing water from draining out of the system. However, the authors argue that full, long-term wetland restoration requires a broader scale of changes done at the watershed level to restore past regime flows. Without proper hydrology, restoration goals for a restored wetland may not be fully met.

The authors propose using an "ecosystem-level perspective" which also includes looking at the various ecosystems that were present in the past. Although restoring forests will still be a crucial part of many restoration projects, in other cases wetland or other ecosystem restoration should take priority. A balance of herbaceous wetland plants along with trees and shrubs and a focus on restoring past water flow regimes will lead to overall improved restoration for a whole watershed.

Instead of reconstructing water flow regimes to benefit riparian forests, a broader approach should be implemented to improve ecosystem mosaics. A mix of ecosystems supports a broader range of species that fit into certain habitats and microclimates. Restoring natural regimes leads to increased land mosaics and requires less active management in the future. Restoring fluvial marshes can be created and managed even regulated river systems by controlling inundation. The amount of flooding these marshes receives leads to a healthy buildup of fine sediments that can prevent siltation father downstream and prevent woody plants from inhabiting an area of ecosystem where they would not be traditionally.

An integral part of this perspective for restoration is understanding and modeling hydrological regimes. Many wetland species developed to grow during certain times within the regimes. Excess water at the wrong time can lead to flooding and prevent herbaceous plant species from reestablishing in their native habitat. Currently, there are no models that exist to predict and manage hydrological schemes to wetland species.

Although these concepts have been noted in the past the authors particularly focus on the important of wetland restoration within a riparian ecosystem. The landscape context is an important aspect of all restoration projects and specially care should be used to prevent inappropriate species from invading areas where they were traditionally absent. Current restoration efforts do little to restore not only forests but various types of wetlands that have been greatly reduced.

Response of Six-Lined Racerunner (*Aspidoscelis sexlineata*) to Habitat Restoration in Fire-Suppressed Longleaf Pine (*Pinus palustris*) Sandhills

Many ecosystems that have traditionally been maintained by fire are threatened by landscape changes and public policies of fire suppression. In some cases, fire-adapted species may not be able to reproduce or other species may encroach on an ecosystem. It can be difficult to restore fire-suppressed ecosystems because there is often an excessive buildup of fuel that can cause a fire to burn hotter than is healthy for restoration. Often, manually removing the excess fuel is necessary, but this act does not substitute for the ecological benefits of fire. Therefore, a combination of the two may be more helpful. Steen *et al.* (2013) used a mark-release strategy to study the health of the ecosystem by looking at the population response for the indicator species *Aspidoscelis sexlineata*. Test sites were separated into ones treated by burning, mechanical removal of brush, herbicide, and control (no fire). Sites were exposed to fire in regular 2-year intervals for a decade. After 4 years, populations increased in both the fire-only and mechanically-treated sites. After 10 years, all sites experienced population growth of *A. sexlineata*. The results indicate that prescribed burning alone is sufficient for ecosystem recovery.

Steen and colleagues used 24 81-ha sites of Longleaf Pine sandhills to conduct their research along with 6 reference sites that represented target ecosystems. The sites were broken up into hardwood removal by burning, removal by herbicides, and removal by mechanical methods. In 1995, burn sites were burned between April-June, herbicides were applied in May, and trees were mechanically removed between June and November. Data for the vegetation was collected in both 1998 and 2009, and re-sampled in 2010. Information was taking about the percent coverage of ground cover of different variables (grass, branches, leaves, etc.) and the number and dbh of Longleaf Pine trees. Pitfall traps were placed at the center of each test site and specimens were analyzed between May to August 1997 and April to August 1998. From May to September 2009 and May to August 2010 pitfall traps were placed at the same locations in addition to box traps. Individual *A. sexlineata* were sorted based on size into juvenile and adult groups. Adults were then sorted based on gender using marking on their skin.

A before-after-control-impact study was used to compare the change of adults and juveniles in the test and reference sites over time. If there was no difference between the populations at test and reference sites then it would be assumed that the restoration effort had reached its target goals. However, the number of squamates (reptiles) marked and recaptured remained low and it was difficult to obtain reasonable estimates of *A. sexlineata* populations. The authors suggest that future studies spread pitfall traps throughout a site rather than in the center, as was done in this study.

The results found that there was no significant change in the populations of *A. sexlineata* between the 1997-1998 sampling and the 2009-2010 sampling. Between 1997 and 1998, the number of individuals at the burn and mechanical sites did not significantly differ from the reference sites. The control and herbicide sites had significantly less squamates compared to the reference sites. Between 2009 and 2010, there was no significant difference between the mean number of individuals at the reference sites and all four test sites.

Long-term prescribed burning increased the mean number of individuals at all sites, regardless of previous fuel removal. The number of individuals observed remained relatively constant over-time and showed no significant changes between 1997-98 and 2009-10 sampling. However, there is a large possibility that small samples sizes prevented statistical analysis from seeing any significant trends between the 1997-98 control and herbicide sites and the same sites from 2009-10. Hardwood trees grew back gradually at all sites except for control sites.

There was a marked increase in *A. sexlineata* in burn sites along with sites with mechanical removal of hardwoods followed by prescribed burn. Over time, all sites had similar squamate populations compared to populations at reference sites. Other studies have observed that simply reestablishing disturbance patterns in an ecosystem may not lead to pre-disturbed species assemblage. However, for this study, reintroducing fire at consistent intervals was enough to help increase the population of an indicator species.

One aspect of the study that was not fully understood was the number of individuals needed for a viable population. The data collected may not reflect effective population size in each of the test sites. The authors did not know the minimum number of individuals that reflect and a viable population and chose to look at the number of *A. sexlineata* as a target condition.

Although it is possible that immigration played some role in the increase of *A. sexlineata* over time, evidence suggests that recruitment and fecundity may have played a larger role. Traps were set in the center of each site and the number of juveniles in test sites was similar to those at reference sites. *A.* sexlineata mature quickly at one year and the number of juveniles most likely reflects reproductive success. The response of the reptile populations was most likely due to changes in vegetation structure. The authors suggest that vegetation structure changes created microclimates more suitable for the reptiles or made it easier to hunt arthropod prey.

It is likely that even with thought the *A. sexlineata* were living in a degraded ecosystem, they were still able to survive, albeit with lower survival and reproductive rates. It should also be noted that hardwoods should not be completely taken out as some species might use or require them within the ecosystem. Burning the ecosystem at consistent intervals was enough to increase the population of the reptiles to a level similar to that of reference sites. For Longleaf pine sandhills, regular burning has a positive effect and can be used in the future to protect *A. sexlineata.*

Conclusions

Ecological restoration is a growing field that will play an important role in the future. Already we have seen ecological benefits in many projects and the economic paybacks from proper restoration may help to push policy-makers to invest in restoring degraded habitat. As we have seen, there are important local and global decisions that need to be made in order to protect biodiversity and protect against environmental destruction. More knowledge is needed in order to successfully restore ecosystems to their pre-disturbed states but our knowledge continues grow rapidly to combat permanent habitat destruction. As time goes on, restoration techniques will continue to develop and become more precise to handle individual ecosystem needs. We cannot reasonably believe that we can set out to restore all of the world's degraded ecosystems, but any habitat restoration that can be done will be beneficial.

References Cited

Aronson, J. and Alexander, S. (2013), Ecosystem Restoration is Now a Global Priority: Time to Roll up our Sleeves. Restoration Ecology 21, 293–296.

Barbier EB, Georgiou IY, Enchelmeyer B, Reed DJ (2013) The Value of Wetlands in Protecting Southeast Louisiana from Hurricane Storm Surges. PLoS ONE 8(3)

Bennartz, R., Shupe, M., Turner, D., Walden, V., Steffen, K., Cox, C., Kulie, M., Miller, N., Pettersen, C., 2013. July 2012 Greenland melt extent enhanced by low-level liquid clouds. Nature 496, 83-86.

Clewell, A. F. and Aronson, J. (2006), Motivations for the Restoration of Ecosystems. Conservation Biology, 20: 420–428

Halme, P *et al. Challenges of ecological restoration: Lessons from forests in Northern Europe.* 2013. Biological Conservation 167, 248-256.

Ilmonen, J., Virtanen, R., Paasivirta, L., Muotka,T. 2013.Detecting Restoration Impacts in Inter-Connected Habitatss: Spring Invertebrate Communities in a Restored Wetland. Ecological Indicators. 30, 165-169.

Shackelford, N., Hobbs, R. J., Burgar, J. M., Erickson, T. E., Fontaine, J. B., Laliberté, E., Ramalho, C. E., Perring, M. P. and Standish, R. J. (2013), Primed for Change: Developing Ecological Restoration for the 21st Century. Restoration Ecology, 21: 297–304.

Steen, D. A., Smith, L. L., Morris, G., Mike Conner, L., Litt, A. R., Pokswinski, S. and Guyer, C. (2013), Response of Six-Lined Racerunner (*Aspidoscelis sexlineata*) to Habitat Restoration in Fire-Suppressed Longleaf Pine (*Pinus palustris*) Sandhills. Restoration Ecology, 21: 457–463.

Throop, William. 2000. *Environmental Restoration: Ethics, Theory, and Practice.* Humanity Books.

Weisberg, P. J., Mortenson, S. G. and Dilts, T. E. (2013), Gallery Forest or Herbaceous Wetland? The Need for Multi-Target Perspectives in Riparian Restoration Planning. Restoration Ecology, 21: 12–16.

16. Climate Change and Extinctions

Elizabeth Medford

Climate change has been predicted to affect the planet in a variety of ways and severity levels. While many discussions regarding the consequences of global warming focus on the immediate affects on human living standards, other important aspects to consider are the affects that climate change will have on all the other organisms that inhabit the planet. The ability to predict the changing populations and distributions of many of Earth's species will prove beneficial for anthropogenic reasons in addition to the intrinsic value of having the knowledge. Species like bees and other pollinators are important for agricultural purposes while marine species are important for fulfilling much of the world's dietary needs. Moreover, this information can be used in the future to implement efficient and meaningful policy regulations relating to vulnerability of species and protecting necessary habitat.

Understanding the particular interactions between population and climate are complicated and currently not well documented. However, using various population modeling techniques, many studies have focused on specific species and have been able to make general conclusions about the ways that different populations might change in response to a warming planet. Generally, responses include changing body size and distribution shifts to higher latitudes, closer to the poles. Looking at past population responses to warming events can also influence the ability of models and researchers to accurately predict the way that species will respond to warming. From the studies analyzed below it becomes apparent that the mechanisms through which climate change will lead to extinction or population decline are more complex than initially predicted. Instead of increasing temperature leading directly to population declines or decreased fitness, it appears that instead, climate will alter interactions between species, which will in turn reduce some populations and enhance others. These interactions however vary from species to species and the amount they will be affected varies greatly. Generally, one can conclude that most species will be able to survive climate change but will have a different habitat than currently or will have decreased genetic diversity increasing the probability of population declines in the future, but not decreasing population immediately.

Elizabeth Medford

How Does Climate Change Cause Extinction?

One concern related to the consequences of anthropogenic climate change is the extinction of vulnerable species. While climate change has been predicted to affect the populations of these species, few studies have concretely identified the precise mechanisms through which populations will be affected. Using population studies and the empirical support for climate related extinctions, Cahill *et al.* (2013) reviewed the many different avenues through which extinction may occur as a result of climate change. By examining both the direct and proximate factors causing extinction, this review outlines each possible cause and presents any known empirical support for each cause. Cahill *et al.* additionally provides instruction on mechanisms for finding proximate causes for extinction. The reviewed data highlight that changes in species interactions are the important proximate cause of extinction relating to global climate change and that further studies will be crucial for developing effective conservation strategies.

Many different causes for extinction relating to changes in the Earth's climate have been proposed, however few have ample empirical evidence due to the complex nature of population dynamics and species interactions. In the past, these proximate causes have included negative impact of heat-avoidance behavior, loss of host and pollinator species, and increased pathogen and competitor populations, among others. In their review, Cahill *et al.* grouped proximate causes into different categories helping to evaluate the validity of each. Categories included: temperature, precipitation, temporal mismatch between species, negative impacts on beneficial species, and positive impacts on harmful species. Once the causes were clearly stated and organized, they could be properly evaluated on scientific validity as causes. The clearest change in species behavior resulting from climate change is a pattern of range shifts documented in hundreds of species. The patterns of warm-edge contraction specifically provide evidence that local extinctions have already occurred as a result of climate change. The online supplementary material for this review includes an extensive list of studies on extinctions occurring in the recent past. Of these 864 global species extinctions, only 20 are considered by the International Union for Conservation of Nature (IUCN) to have resulted from climate change. Specifically, Cahill *et al.* referenced coral bleaching and chytrid fungus in amphibians as two examples of climate-related extinctions. Of these 20 extinctions, seven were amphibian species, four were snails, two fish species, six bird species, and one rodent. Interestingly, none of these twenty became extinct as a result of limited tolerances to high temperatures.

The authors found that climate change will cause extinctions mostly by changing the way species interact with each other, which will affect food availability and breeding patterns. Surprisingly, species' ability to adjust to higher body temperatures has not been the prominent proximal cause of global extinction. This review urges for further studies on the effects of climate change on species interaction and extinctions in general. Biophysical modeling and population range surveys are the two methods suggested by Cahill *et al.* for increasing the amount of data on populations that may be affected by the negative results of climate change. These additional studies would to help improve the effectiveness of conservation strategies, which will undoubtedly become more critical as climate change progresses and the number of related extinctions continues to rise.

The Shaping of Genetic Variation in the Edge-of-Range Populations under Past and Future Climate Change

As the climate continues to warm because of human influences on the chemical composition of the atmosphere, organisms will need to adapt to new conditions or move to new habitats. While many species have proven able to adjust their ranges when faced with changes in temperature or food availability, the consequences of range-shift have not been widely studied. Razgour *et al.* studied a species of bat (*Plecotus austriacus*) in the hopes of better understanding the decreases in genetic diversity that can occur after range shifting caused by climate change. The authors extracted genomic DNA from 259 individuals and conducted various genetic analysis tests on the DNA including PCR. Additional data included genetic diversity information from models of past climate-related range shifts in Europe. The data were combined to predict future declines in genetic diversity for *P. austriacus* populations in Europe. The study provides evidence that geographical barriers shape genetic variation and that in the future, genetic diversity in *P. austriacus* will be reduced by more than half, as it will for many other temperate European species.

Future changes in temperature and precipitation will not only affect the population sizes of species but may also affect the genetic diversity of different communities by killing off all but a select few individuals in a particular habitat. Studying the genetic diversity of current bat populations can provide insight into how past climate events have shaped genetic diversity. Razgour *et al.* constructed phylogenetic trees and utilized numerous genetic analysis programs to determine Iberia as the location of the most diverse populations of bats with the greatest number of unique haplotypes and private alleles. This is likely the origin of the species after the Last Glacial Maximum (LGM), which cause range shifts and reduced genetic diversity in the species.

Evidence of past destruction of genetic diversity can be found in fossil records from the past climate changes in Europe following the LGM and receding of the glaciers. Ecological niche modeling and ABC inference of demographic history models were both used to construct a hypothetical range for the bats in the past and future. These studies predicted that most of the current habitat range for the bats would no longer be suitable habitats by 2080. This range restriction will cause more than half of the species' genetic diversity to be located in unsuitable locations, which will result in vast diversity losses. Additionally, limited contemporary gene flow across the Pyrenees Mountains suggests that geographic barriers like mountains will further restrict the range of *P. austriacus*.

Razgour *et al.* studied *P. austriacus* for the species' diversity, wide distribution, keystone ecological roles, and sensitivity to changes in temperature. Besides these attributes, bats also have a longevity and reproductive rate that suggest that they may not be able to evolve quickly enough for future changes in temperature and precipitation. The study concludes that the future for these bats will include northern range expansion and southern range contraction by the end of the century. These range shifts will likely cause the most extensive loss of genetic diversity in the most diverse populations. Moreover, this loss of genetic diversity may make the bats more vulnerable to threats like disease and challenges relating to more severe changes climate. Therefore, the authors of this study urge further research into the effects of climate change on leading-edge populations, as they will play a significant role in range shifts and spreading genetic diversity.

The Past, Present and Potential Future Distributions of Cold-Adapted Bird Species

Population distribution models play an important role in analyzing past changes in species ranges and in projecting future adaptations in different species. Specifically, bird species provide important data regarding distribution adaptations because of they can more rapidly adjust to changes in climate than can non-flying species. Experimentally validating the accuracy of distributional models becomes important when such models are used to project future changes and such projections are used for policy decisions. Smith *et al.* 2013 tested the accuracy of climate-only distributional models by comparing model projections for four species of birds after the last glacial maximum (LGM) with the fossil records of the same species. These comparisons proved that two of the species of birds were in fact located at lower elevations during the LGM than they are today. Moreover, from this study it becomes evident that in the future, changes in climate may lead to the intersection of these bird species. From this study the authors conclude that climate-only projection models can accurately predict distribution changes.

Since climate is not the only factor acting on species distribution, often climate-only models receive criticism that they do not holistically analyze reactions to climate change. The second-generation models focus not only on climate but also on variables such as habitat availability, dispersal capacity, population growth rates, and species interactions. One way to test whether the first generation climate-only models can play any role in projecting future distributions is to test their accuracy compared to known distribution changes. Smith *et al.* chose to test these models by comparing the distribution changes modeled for four bird species after the LGM with the known distribution changes from fossil records. These species included the Willow grouse *Lagopus lagopus*, Rock ptarmigan *Lagopus mutus,* Yellow-billed chough *Pyrrhocorax graculus,* and the White-winged snowfinch *Montifringilla nivalis.* The authors choose these species for their adaptation to cold arctic climates and for the fact that they may be particularly threatened by future warming. Additionally, the fossil records available for these species are exceptional in that they can be georeferenced, dated, and identified on a species level.

To conduct their analysis the authors compiled data from the European Bird Census Council as well as European listings of archeological dig locations with fossil and sub-fossil records for each bird species. Climate data were collected from the Worldclim database using data from the period 1950–2000. Smith *et al.* utilized climate models from the Community Climate System model version 3 and the Model for Interdisciplinary Research On Climate version 3.2 to produce global climate simulations for the LGM. By combining the information from all of these sources the authors found that each model accurately predicted the current distributions of each species. LGM projections show different past distributions for the four species with *L. lagopus* occurring further south than currently and *L. mutus* occurring more widely in southern Europe. These models were also used to project future changes including the further latitudinal retreat caused by anthropogenic warming for *L. lagopus* and *L. mutus.*

From this study, Smith *et al.* conclude that climate is likely a key driving force for distributional changes for these cold-adapted bird species. The fossil records and the LGM-projected data matched reasonably but not perfectly indicating needed improvements in modeling technology. More generally, this study implies that even with closely related bird species there are individualistic responses to changes in climate. For example, some species shift the latitude of their ranges while other species shift the elevation at which they live. Projections of the future distributions of these four bird species indicate that they may co-occur again in the Scandinavian mountains if the climate continues to warm. Therefore, climate-only models can be used to accurately predict species distribution despite the worry that these models fail to consider all components involved in distribution changes.

Identifying the World's Species Most Vulnerable to Climate Change: A Systematic Trait-Based Assessment of all Birds, Amphibians and Corals

While it has been recognized in the past that climate change will have impacts on biodiversity, many approaches ignore the differences between species that will increase or reduce their vulnerability. Foden *et al.* (2013) chose to address three different aspects of climate change vulnerability to account for species' biological traits: sensitivity, exposure, and adaptive capacity. In combining these traits with the modeled exposure to projected climate change, the authors assessed the species with the greatest relative vulnerability to climate change. These methods were applied to each of the world's birds, amphibians, and corals. The authors also identified the geographic areas in which the most vulnerable species are concentrated. These included the Amazon basin for amphibians and birds, and the Indo-west Pacific for corals. The aim of Foden *et al.* is that the results from this study will help to better protect vulnerable species from the dangers of climate change. The data from this study can be used to devise more effective species and area specific conservation strategies.

Recently, anthropogenic climate change has become an official significant threat for vertebrate populations according to the Intergovernmental Panel on Climate Change (IPCC). The studies done by the IPCC however have been focused mostly on global assessments of potential climate change impacts. To more accurately identify the species most at risk from climate change Foden *et al.* incorporated species' physiological, ecological and evolutionary characteristics with predicted climate change exposure. To identify the biological traits that determine species' climate change vulnerability the authors held two workshops with over 30 experts in extinction risk over a broad range of taxonomic groups. These workshops resulted in over 90 biological, ecological, and environmental traits likely to influence climate change vulnerability. After consolidating these traits the authors finalized 'trait sets' including habitat specialization, narrow environmental tolerances, the potential for disruption of both environmental triggers and interspecific interactions, rarity, poor dispersal potential, and poor micro-environmental potential due to low genetic diversity. The authors study nearly 10,000 bird species, 6,000 amphibian species, and 800 warm-water reef-building corals because these taxonomic groups are well studied and include species from terrestrial, freshwater and marine biomes. Much of the trait data was gathered using online databases, experts' knowledge and the International Union for Conservation of Nature's (IUCN) Red List.

The results from this study assess relative vulnerability rather than absolute vulnerability to climate change, meaning that the results cannot be used to infer how many species will be impacted, nor vulnerability between taxonomic groups. Instead, the results can best be utilized to infer which species are likely to be at the greatest risk of extinction driven by climate change. The authors included both the total numbers of vulnerable species in a region and their proportion relative to all species occurring, in the hopes that areas containing the greatest number of highly climate change vulnerable species will be protected. The Amazon for example emerges as a region of utmost importance because of the high concentration of climate change vulnerable bird and amphibian species. Large numbers of highly vulnerable bird and amphibian species also can be found in Mesoamerica. For corals, the proportion of highly vulnerable species shows little spatial pattern, but with a slight concentration of such species in the Caribbean. The information garnered by Foden *et al.* is vital for large-scale conservation planning exercises but also illuminates that more detailed assessments are needed to fully understand climate change vulnerability.

Temperature Drives the Continental-Scale Distribution of Key Microbes in Topsoil Communities

While it has become fairly well known that global warming will cause plant and animal species to migrate toward cooler areas or cause range loses, until now it has been unclear whether this will also be true for microorganisms. Microorganisms play a key role in soil fertility and erodibility making this study relevant both for future agricultural endeavors as well as future efforts relating to ecological protection. Garcia-Pitchel *et al.* (2013) conducted continental-scale compositional surveys of soil crust microbial communities in the arid regions of North America. The results from these surveys imply that temperature caused latitudinal replacement between two key topsoil cyanobacteria. The cyanobacteria *Microcoleus vaginatus* behaved more psychrotolerant and less thermotolerant while *M. steenstrupii* behaved more thermotolerant. These results imply that the later may replace the former as temperature increases globally. Further studies are required to fully understand the impact of this microbial replacement.

Soil crusts, also known as biocrusts, are largely microbial communities that play crucial roles in soil management for plant interspaces in arid lands. Some of the roles of these photosynthetic assemblages include helping to stabilize the soil against erosion, modifying the hydrological properties of soil, and exporting biologically fixed carbon and nitrogen. These contributions to the desert ecosystem are important globally and locally because of the extent of arid lands. Cyanobacteria were chosen for this study because of the abundant prior descriptive work available on these bacteria that make up the biocrust's dominant phylum. The two cyanobacteria, *M. vaginatus* and *M. steenstrupii*, dominated the phototroph community, which despite having similar generic names are not closely related phylogenetically. Despite these differences, both species are rope-formers meaning they help to stabilize soil on contact and they act as biocrust pioneers. Garcia-Pichel *et al.* also choose these species because of their role as keystone species in desert communities.

To evaluate the ranges of the two chosen cyanobacteria the authors performed a continental-scale survey of bacterial diversity using 16*S* rRNA gene diversity in community DNA. They additionally evaluated each site on soil type, geochemistry, texture, geography, and climate. Simple correlation and multiple regression analyses applied to the distribution of individual taxa helped to determine which cyanobacteria abundances correlated best with mean annual temperature. To test the temperature segregation hypothesis, the authors used two cultivation avenues, which included observing enrichment cultures at different incubated temperatures. The results from these incubations indicated that there were differential responses to high temperature between the two species. None of the *M. vaginatus* culture survived incubation at 40°C while *M. steenstrupii* strains performed well at 35°C. Conversely, at low temperatures (10°C) as well as in the psychrophylic range, *M. vaginatus* strains grew more than the *M. steenstrupii* strains. This experimental data as well as the geographic distribution imply that *M. vaginatus* represents a more psychrotolerant taxon while *M. steenstrupii* is a more thermotolerant family. With this information Garcia-Pichel *et al.* can logically predict that a few degrees of temperature increase caused by anthropogenic climate change can cause a replacement in the dominance of *M. vaginatus by M. steenstrupii* especially on the cooler side of the current boundary. According to climate models, the arid southwestern parts of the United States are among the most likely to see temperature increases in the order of one degree per decade. Therefore, in fifty years all of the *M. vaginatus* will be replaced with mostly unknown consequences for soil management and ecological restoration efforts in arid landscapes.

Diversity Loss With Persistent Human Disturbance Increases Vulnerability to Ecosystem Collapse

The value of biodiversity for the resilience of ecological systems has become common knowledge in ecology spheres. Moreover, the affects of humans on the diversity of many ecosystems around the world have been proven previously. This study connects these two pieces of knowledge in stating that the combination of diversity loss caused by human activities and environmental change increase the risk for sudden ecosystem collapse. MacDougall *et al.* (2013) demonstrated this connection in a degraded but species-rich grassland that was subject to fire suppression techniques as well as invasion by non-native species. The authors conclude that human disturbance created a negative relationship between diversity and function but that the elimination of the buffering effects of high species diversity has led to a vulnerability to sudden environmental change. The findings of this study can be applied more generally to many different ecosystems because of the prevalent combination of long-term land management and species loss. This study relates to climate change because as global temperatures cause fires to increase in occurrence and severity, trophic collapse because of human-caused diversity loss may also increase.

Since the mid-nineteenth century many forest communities have been managed similar to agricultural crops. Fire-suppression has been an integral part of such forest management. While these fire-suppression practices have long been associated with loss of plant species, communities also have a relatively high annual productivity level and invasion resistance compared to other managed low-diversity agricultural systems. To test the vulnerability of managed grassland areas, the authors established 80 meter sized treatment plots on Vancouver Island, British Columbia, Canada. This savanna has 115 plant species, 75% of which are native and include grasses, forbs, and woody plants. Biomass was measured in half of the plots annually and nutrient levels were measured in 2000 and two years after burning.

The authors burned twenty of the plots once a year for ten years while twenty other plots were burned similarly but only for five years.

Redundancy can be crucial for stability in complex systems because one species can substitute the function and niche of another. In the case of the human influenced grassland however, no species was distributed widely enough beforehand to compensate solely for the effects of fire. MacDougall *et al.* observed that biodiversity levels not only influenced invasion after burning but also the dynamics of the fire regime itself. With the re-introduction of fire the grasslands only survived in areas with remnant native species where rare plants prevent extensive invasion and conversion toward woodland. Therefore, the system collapsed within one growing season with immediate dominance by invasive species. This study illustrates the ways that human activity can reduce biodiversity levels and therefore homogenize the structure and function of ecological systems. We can see from this study that this homogenization simultaneously weakens compensatory mechanisms for responding to sudden changes. Considering that human activity has been persistent in many different ecosystems around the world, there may be similar vulnerability to sudden environmental change. Unfortunately, this vulnerability will only be evident after the collapse occurs making preventative action difficult. Considering the environmental disturbances predicted to occur from changing temperature and precipitation levels, these collapses may occur more frequently than they are currently, making ecosystem vulnerability even more of a pressing topic. The authors urge for further studies on the effects of human disturbance in other ecosystem types to better understand possible future system collapse.

Exploring Tree Species Colonization Potential using a Spatially Explicit Simulation Model: Implications for Four Oaks under Climate Change

While the impact of climate change on a variety of animal populations and their ranges has been studies extensively in the past, the study of the impact of warming on tree species also provides useful information for policymaking. A variety of different modeling systems apply different variables and make predictions about tree species distribution in the future as temperatures rise. In this study however, Prasad *et al.* (2013) combine two different commonly used technologies to overcome the constraint of computation time and allow assessment of colonization potential for oak species. Four oak species were chosen to focus on because they are strongly climate-driven species: black oak, post oak, chestnut oak, and white oak. Using the DISTRIB and SHIFT models together the authors were able to determine the future dominant forest types in the northeastern United States. This study determined that even under optimistic conditions ignoring some influential factors, only a small fraction of suitable oak habitat is likely to be occupied by oaks within 100 years. The authors urge that the information garnered in this study be used to inform assisted migration practices for vulnerable tree species. They additionally call for further studies focusing on how each individual species will adapt to increases in temperature.

Trees have three different options for responding to climate change; they can adapt, migrate, or go extinct. Luckily the high genetic variability in many tree species allows for rapid adaptation to environmental changes. This adaptation ability can be traced to phenotypic plasticity and abundant fecundity, both of which help to explain their ecological success. It is important to understand the effects of climate change on tree species because of their exposure to changing environments and interspecific interactions. However, genetic constraints on adaptation and impediment to both gene flow and dispersal by landscape fragmentation could threaten tree species in the era of climate change. To evaluate the effects of climate change on four different oak species the authors used SHIFT modeling technology along with DISTRIB technology. SHIFT models use historical tree migration rates to estimate future colonization with light parameters applied to the model. The DISTRIB model is used to predict and map current and future-climate tree habitats for 134 tree species via changes in relative importance values. Combining these two modeling systems the authors were able to use historical information on rates of past migration events to predict future potential migration. This new approach accounts for the structure of the landscape using information regarding fragmentation of habitat quality, information that many models ignore. The conclusions obtained from the combination of these models can be used for protecting vulnerable species in the future.

The results of these studies imply that minimal amounts of the potentially colonizable habitat would actually be colonized, highlighting the dangers of relying on models that provide information about the potential colonization by tree species. Specifically, Prasad *et al.* found that only 5% of suitable habitat would be colonized and that only 2% of habitats have a 50% chance of being colonized. Moreover, the authors ignored additional factors that could play a role in limiting colonization by trees such as insects, disease, and competition from maple species. Studies that included these other factors would likely produce even lower rates of predicted colonization by oak species. However, the authors also ignored the characteristics of the oak trees that could potentially benefit them in a warming climate, such as their drought tolerance. The findings of this study can be used to initiate new populations, and to facilitate migration, and these populations do not need to be large. Specifically, these studies can be used for specialist species that have narrow niches, which will likely be more negatively affected by climate change. Additionally, the DISTRIB and SHIFT models provide information about suitable corridors and patches for facilitating migration of tree species. For example, one migration corridor for the black oak is in Wisconsin. In the future this information can be used for facilitated migration of black oak trees.

This study is significant because of the way that the authors applied different models to overcome the computational barrier and tackle many species efficiently. Moreover, Prasad *et al.* were able to account for multiple scenarios for each species, which will be essential for understanding the affects of global warming on the distributions of different tree species.

The Future of Species Under Climate Change: Resilience or Decline?

While it is widely accepted that as climate changes species will be affected, to what extent, and which will be most vulnerable are both still largely unknown. Looking at past changes in climate and the corresponding fossil records illuminates the ability of many species to adapt and persist through variations in climate. However, models predicting future ranges and populations for many species forecast large-scale range reduction and extinction. To reconcile these two differing arguments about the ways that species will be affected by climate change, Moritz *et al.* (2013) summarized salient concepts and theories, and reviewed various predictions about future response and evidence from paleontological studies of past responses. They conclude that the most likely response to climate change will be decreasing body size and geographical range shifts rather than extinctions, but that further research is required for understanding the capacity of species to buffer effects of climate change.

Understanding the variability of different species is essential for forming future conservation policies. Generally, the authors argue that the vulnerability of a species is a combination of exposure to changes in climate and intrinsic sensitivity relating to physiological limits or trophic specialization. These factors combine to determine whether or not a species will be able to buffer climactic alterations. Exposure includes not only changes in temperature but also increases in frequency of extreme climate events, which will also affect species persistence. Additionally, landscape features such as slope, aspect, vegetation cover, and soil moisture can ameliorate shifts in extreme temperatures. In response to exposure to these and other changing environmental conditions, species can alter fitness-related traits by plastic change or genetic adaptation. However, Moritz *et al.* mention that so far, direct evidence of genetically based adaptation to climate change over time remains sparse indicating that genetic response is not generally a viable option for species. Therefore, the most viable response is local shifts in microhabitat use and dispersal. Because of the complexities associated with changing ranges and species interactions, understanding this response requires more studies.

Currently, population decline and extinction predictions formulated from models vary greatly in how severe they claim declines will be. For example, one prominent analysis predicted that 15–37% of species would be endangered or extinct by 2050 while another predicts more than a 50% loss of climatically suitable range by 2080 for 57% of plant species and 34% of animals. Generally, montane-taxa are predicted to lose range area and shift upward in latitude. However clearly there are grave uncertainties and inconsistencies in future predictions. This led the authors to look back on fossil records to see how species responded to past shifts in global temperature, specifically at the rapid warming events and the Pleistocene-Holocene transition which implied that there was no signal of elevated extinction though periods of rapid change and the trophic structure of mammalian communities appear robust. However, there were also more recent megafaunal extinctions, in which climate change and human impacts likely combined with devastating consequences. Fossil records indicate that the most common responses to past climate change were range shifts and decreases in body size. The authors acknowledge however that the discord between predictions of high extinction and high resilience though paleoclimatic change might be hidden by limitations in the fossil record.

Conclusions

In general, much more research on the specific responses to different species and taxa will be necessary to understand the varying responses to climate change. Specifically, improving model-predicting technologies will vastly improve the discipline of range shift prediction. Along with these improvements, increasing past and present population data for many species will help to test the improved accuracy of many models. Despite the many missing studies that would make concrete conclusions about future populations more reliable, some general conclusions can be made based on the articles reviewed here. Many authors conclude in their studies that likely responses to a variety of species will include range shifts to higher latitudes and decreases in body size. Additionally, for many species, range shift will decrease genetic diversity, which has the potential to increase vulnerability to extinction. Combining further studies with the information garnered from these studies, more effective and precise policy decisions can be made to preserve vulnerable species from the consequences of climate change.

References Cited

Cahill, A., Aiello-Lammens, M., Fisher-Reid, M., 2013. How does climate change cause extinction? Proceedings of The Royal Society Proc R Soc B 280: 20121890

Foden, W., Butchart, S., Stuart, S., 2013. Identifying the World's Most Climate Change Vulnerable Species: A Systematic Trait-Based Assessment of all Birds, Amphibians and Corals. Plosone 8.

Garcia-Pichel, F., Loza, V., Marusenko, Y., 2013. Temperature Drives the Continental-Scale Distribution of Key Microbes in Topsoil Communities. Science 340, pp. 1574–1577.

MacDougall, A., McCann, K., Gellner, G., 2013. Diversity Loss With Persistent Human Disturbance Increases Vulnerability to Ecosystem Collapse. Nature 494, 86–89.

Moritz, C., Agudo, R., 2013. The Future of Species Under Climate Change: Resilience or Decline. Science 341, 504–508.

Prasad, A., Gardiner, J., Iverson, L., 2013. Exploring Tree Species Colonization Potentials Using a Spatially Explicit Simulation Model: Implications for Four Oaks Under Climate Change. Global Change Biology 19, 2196–2208.

Razgour, O., Juste, J., Ibanez, C., Kiefer, A., 2013. The shaping of genetic variation in edge-of- range populations under past and future climate change. Ecology Letters doi 10.111/ele. 12158

Smith, S., Gregory, R., Anderson, B., 2013. The past, present, and potential future distributions of cold-adapted bird species. The Journal of Conservation Biology DOI 10.111/ddi. 12025.

17. Extinction Risk and Climate Change

Samantha Thompson

Our planet is undergoing a biodiversity crisis. Globally, at least 16,000 species are threatened with extinction, including 12% of birds, 23% of mammals and 32% of amphibians (Suzuki, 2013). Biologists know what is causing this environmental crisis—human impacts from development, deforestation, pollution and climate change are destroying the homes and habitat of wildlife around the world. More importantly, biologists understand that the trend can be reversed. There was a time when populations of the great whales, bald eagles and whooping cranes were in rapid decline. But because of strong legislation, habitat protection and international agreements, these populations are bouncing back. For the majority of species that make the list, the government has chronically failed to identify and protect the habitat they need to survive and recover. This thus causes species to become threatened, endangered, and even go extinct.

Biodiversity is best defined as the variety of life (its ecosystems, populations, species and genes). Saving endangered plants and animals from becoming extinct and protecting their wild places is crucial for our health and the future of our children. Biodiversity is also important because its loss affects many services that are essential to the functioning of our society and economy.

As species are lost so too are our options for future discovery and advancement. The impacts of biodiversity loss include clearly into fewer new medicines, greater vulnerability to natural disasters and greater effects from global warming (Memije-Cruz, 2010).

The main threats to biodiversity arise from loss of habitat and disruption of ecosystems due to pressure of growing populations and unsustainable resource use. The most important causes of extinction related to human activity involve habitat alteration for clearing and burning of forests, exploitative and illegal logging, and fuel wood collection. Encroachment, wetland conversion, degradation of grasslands, inappropriate development in coastal areas and urbanization also significantly impact biodiversity loss which causes species to become endangered. Poaching, hunting, introduction of exotic species, and pollution also contribute substantially to biodiversity loss. Marine life is also vulnerable to outside intervention such as extensive oil explorations. By becoming aware of the impact we have on the environment, humans can help prevent biodiversity loss.

Resilience of Marine Turtle Regional Management Units to Climate Change

Scientists are searching for ways to avoid losing biodiversity to global warming. One way is by enhancing species resilience to environmental change. Resilience is the ability of an ecosystem or species to maintain key functions and processes in the face of stresses or pressures, either by resisting and/or adapting to change. Fuentes explored the resilience of 58 marine regional turtle management units (RMUs) to climate change, including all species of marine turtles worldwide. Using expert opinions from 33 different IUCN-SSC Marine Turtle Specialist Group respondents, the researchers were able to develop a Resilience Index. This was used in order to consider qualitative characteristics of RMUs such as population size, rookery vulnerability, and genetic diversity, and non climate related threats, such as fisheries, take, and coastal development. Through this information researchers were able to identify the world's 13 least resilient marine turtle RMUs to climate change.

In response to the anticipated widespread negative effects of climate change, research has turned to primarily focusing on strategies that help alleviate some of the effects of the potential threats including the following: 1) mitigate the threat by reducing global green house emissions; 2) adaptively manage impacts from climate change by increasing population persistence; and 3) employ actions that build biodiversity resilience, such as addressing current non climate threats. Because reducing emissions is a big challenge and because immediate reductions will not necessarily stop the already unavoidable effects of climate change, researchers believe it is best to prioritize further research on attempting to predict the impacts of global warming until further information on the effectiveness of the above strategies is obtained. Thus an example of this future type of research can be viewed by further estimating how resilient populations or species can be under future scenarios such as that presented by Mariana Fuentes (2013) and her coworkers.

Experts were brought from several locations worldwide in order to assess each species of turtle used to calculate the RMUs for the resilience index. In some locations only one expert respondent assessed the RMUs of specific species of turtles given their natural habitat location. The RMU resilience index values ranged from a low of 0.89 (most resilient) to a high of 2.08 (least resilient). The variables most likely to influence the resilience of RMUs to climate change were rookery vulnerability and high climate threats. In other words species that suffered from the likelihood of extirpation of functional rookeries would not be able to recover from global climate change and those who suffered from non climate related threats such as fisheries would effect the resilience of RMUs to climate change as well. This can be used as an indicator of persistence of viable nesting in an RMU, given various threats and potential for range shifts over time. This is important because the importance of rookery persistence relates to the need for optimal nesting areas necessary for reproduction and thus recruitment entry into the population.

The ability to predict resilience among a species can help researchers find ways to protect species against environmental factors such as global warming. Because climate change is a major threat to biodiversity it is important that species find ways to overcome natural hardships in order to prevent entire groups of species from becoming endangered or even going extinct. The authors of this essay using an RMU resilience index were able to predict which species of marine turtles would suffer the most due to global warming. With the help of further more research humans can learn ways to help protect species as well. This is important if we want to find ways for species, including ourselves, to overcome the effects of global warming.

The Future of Species Under Climate Change: Resilience or Decline

Species have widely been affected by changes in global climate but to what extent is uncertain, though predictions of species decline are often urgent. For example, one prominent analysis predicted that 15 to 37% of species would be endangered or extinct by 2050 (Moritz, 2013). Another predicts more than a 50% loss of climatic range by 2080 for some 57% of widespread species of plants and 34% of animals (Moritz, 2013). Montane taxa are expected to lose range area as they shift northward with warming (Moritz, 2013). Craig Moritz et al. point out that fossil records suggests that most species have persisted through past climate change, whereas forecasts of future impacts predict large-scale range reduction and extinction. Moritz et. al. explore the apparent contradiction between observed past and predicted future species responses summarizing salient concepts and theories and by reviewing broad-scale predictions of future response and evidence from paleontological and phylogeographic studies of past responses at millennial or greater time scales. Bringing the two ideas together, the authors consider evidence for species responses to recent twentieth century climate changes and place them in a management context.

The vulnerability of a given species to climate change can be viewed as a combination of exposure and intrinsic sensitivity. These factors are mediated by the capacity of local populations to buffer climatic alterations in situ via plastic reactions (including behavioral responses) or genetic adaptation, or by shifting geographically to track optimal conditions. Exposure is typically measured as shifts in mean precipitation or temperature at the mesoscale (e.g., 1 to 100 km^2).

Plastic responses are undoubtedly important for short-term persistence but they can also entail costs and may be insufficient to avoid extinction. Evolutionary rescue requires moderate-to-high heritability of key traits or high potential growth rates of populations, with critical adaptations tracking the rate of climate change.

Forecasts of potential species responses to future climate change come largely in two forms: correlative mechanistic models of individual species, and prediction of higher-level properties such as species richness or turnover. Correlative models are currently the most widespread and scalable method, but they have inherent limits. These models typically apply some form of climate envelope approach, assessing whether the (realized) climate niche occupied by a species continues to exist within the current geographic range and whether it will shift elsewhere or cease to exist. Correlative models are probably a better measure of exposure than of species vulnerability to climate change.

The fossil record and the imprint of history in geographic patterns of DNA diversity (phylogeography) provide information that can be correlated with how species responded to past shifts in global temperature. These sources of information on historical responses have distinct limitations that can be partially overcome by combining types of evidence. The fossil record varies in extent and resolution according to preservation conditions. Phylogeographic analysis, on the other hand, affords higher spatial resolution but typically has low temporal precision compared with fossils.

Comparative phylogeographic studies, often combined with paleoclimatic modeling of geographic ranges, offer another window on past species responses and can identify regions in which taxa persisted through past climate change. When combined with fossil evidence and spatial models, such studies highlight the extent of range shifts but also the importance of scattered microrefugia, which are important for range recovery and perhaps also harbor distinct adaptations. Going further, direct DNA analyses of subfossils provides a much clearer picture of population dynamics through climate change and, for megafauna, highlight differences among species in response to the twin challenges of climate change and human colonization.

The discord between predictions of high extinction under future climate change and relatively high resilience through paleoclimatic change could be partly due to the limitations of the fossil record, but may also reflect the fact that species were previously able to persist in the absence of human-caused impacts on natural systems. In historic times, even though the rate of expected future change may be much faster than that over the past century, there is value in examining how species have responded to climate change over the 20th century.

Shifts in phenology are widely observed in the 20th-century record and could cause temporal mismatch between strongly interacting species, especially where these species employ different environmental cues. As expected with warming, decreasing body size has been observed in several studies of birds and mammals. This response seems to be plastic rather than genetic, or it may be related to extended food availability rather than direct physiological effects.

As yet, no species extinctions are clearly attributable to climate change per se, although several studies recorded local extinctions and population declines. Nevertheless, it is very difficult to establish causative relationships between warming and population declines or extinction, due to the interaction with other anthropogenic factors such as habitat loss or previously unseen pathogens. A recurring message is that we have insufficient knowledge of the proximate cause(s) of observed species declines under global warming: The few examples appear to be more closely related to indirect ecological effects than to demonstrable physiological challenges.

Mutualism Disruption Threatens Global Plant Biodiversity: A Systematic Review

Current species losses are reaching mass extinction levels. As species disappear from an ecosystem, the roles they play are lost, as well. Even if they are otherwise resilient to anthropogenic change, extant species may be affected by disrupted interspecific interactions such as, losing their prey, predators, competitors, parasites, or mutualists. Loss of ecological interactions can impact a wide array of ecosystem processes. It has been argued that every species on Earth participates in one or more mutualisms. As such, broken ecological interactions may by themselves impact global biodiversity, by threatening species that are otherwise unaffected by major drivers of environmental change such as habitat loss, climate change, biological invasions, and overexploitation. Such species are thus vulnerable because their partners are vulnerable, not because they themselves respond directly to broad scale environmental change The reproductive success of most flowering plant (angiosperm) species depends either wholly or partly on animal mutualists providing pollination, dispersal, or seed processing. Animal extinctions disrupting these relationships may create "widow" plant species that exhibit reduced fitness and are vulnerable to coextinction. Using systematic review and synthesis of data from a wide diversity of studies, the authors were able to develop rough but quantitative estimates of the global number of angiosperm species facing widowhood due to the extinction of reproductive mutualist vertebrates. They were also able to estimate the likely impact of widowhood on plant reproductive success. Finally Aslan and cowriters combined the following estimates into their results: the global number of vertebrate-pollinated and vertebrate-dispersed angiosperms; the average number of vertebrate partners per plant species, both globally and by geographic region; the proportion of such partners that are currently threatened with extinction, both globally and by geographic region; and the average decline in reproductive success that widows are likely to experience.

In order to approximate the global number of vertebrate-pollinated and vertebrate-dispersed plants, the following published estimates were gathered and combined: total global angiosperm species richness, the proportion of angiosperms that are animal-dispersed minus those that are ant-dispersed, and the number of genera that are vertebrate-pollinated. Known instances of complete disruption of diffuse mutualisms have occurred on oceanic islands where the number of partners per plant is lower and partners more threatened than on continents. Therefore, the authors examined island and continental values separately. They estimated the number of vertebrate-pollinated and vertebrate-dispersed plants that are island endemics by deriving the percentage of total angiosperms that are island endemics from regional estimates of plant diversity.

Through the review the authors concluded that Africa, Asia, the Caribbean, and global oceanic islands are geographic regions at particular risk of disruption of these mutualisms; within these regions, percentages of plant species likely affected range from 2.1–4.5%. Widowed plants are likely to experience reproductive declines of 40–58%, potentially threatening their persistence in the context of other global change stresses. The systematic approach demonstrates that thousands of species may be impacted by disruption in one class of mutualisms, but extinctions will likely disrupt other mutualisms, as well. Although uncertainty is high, there is evidence that mutualism disruption directly threatens significant biodiversity in some geographic regions. Conservation measures with explicit focus on mutualistic functions could be necessary to bolster populations of widowed species and maintain ecosystem functions.

While uncertainties are large, this extensive research leads the authors to a conclusion that extinctions of currently-threatened animal species may remove all pollination or seed dispersal services from thousands of angiosperm species, globally. As additional animals become threatened over time, the number of plant species affected will also rise as well.

Assessing Global Marine Biodiversity Status within a Coupled Socio-ecological Perspective

People value the existence of a variety of marine species and habitats even though they are negatively impacted by human activities. The Convention on Biological Diversity and other international and national policy agreements have set broad goals for reducing the rate of biodiversity loss. However, efforts to conserve biodiversity cannot be effective without comprehensive metrics both to assess progress towards meeting conservation goals and to account for measures that reduce pressures so that positive actions are encouraged. In order to further asses this progress, the authors developed an index based on a global assessment of the condition of marine biodiversity. They developed this index by using publicly available data to estimate the condition of species and habitats within 151 coastal countries. The assessment also included data on social and ecological pressures on biodiversity as well as variables that indicated whether good governance is in place to reduce them. Thus, the index was a social as well as ecological measure of the current and likely future status of biodiversity. As part of the authors' analyses, they set explicit reference points or targets that provide benchmarks for success and allowed for comparative assessment of current conditions. Overall country-level scores ranged from 43 to 95 on a scale of 1 to 100, but countries that scored high for species did not necessarily score high for habitats. Although most current status scores were relatively high, likely future status scores for biodiversity were much lower in most countries due to negative trends for both species and habitats. The authors also found a strong positive relationship between the Human Development Index and resilience measures that could promote greater sustainability by reducing pressures. This relationship suggests that many developing countries lack effective governance, further jeopardizing their ability to maintain species and habitats in the future.

People appreciate the variety of species in the oceans for their beauty and uniqueness, the natural systems that they collectively create, and the ecosystem services that they support. Even when there is no direct use by humans for them, species and the ecosystems they help build are widely valued for their existence. Because species have aesthetic, spiritual, educational, and scientific value, their loss can generate an enormous amount of public concern. Biodiversity declines have already motivated the dedication of substantial resources toward protecting and restoring species. However, efforts to conserve biodiversity cannot be effective without a framework to assess current extinction risk, level of pressures, and governance factors that can measure progress towards meeting conservation goals.

In this study the authors developed a global assessment of the condition of marine biodiversity using publicly available data to estimate how countries are doing not only in terms of preventing marine species extinctions, but also how countries are preserving the natural marine habitats on which many species depend. Several indices have been valuable for tracking how marine biodiversity is faring at various spatial, temporal, and taxonomic resolutions. Although species diversity indices and assessments have historically focused on terrestrial ecosystems, efforts to quantify and assess marine biodiversity have increased in the last decade (Selig, et. al,. 2013). Red List assessments for terrestrial species began in the 1960s; though some marine species were assessed in the 1990s, the Global Marine Species Assessment (GMSA, http://sci.odu.edu/gmsa) substantially increased that effort beginning in 2005. Although different indices vary in their metrics, they all have an implicit target of at least no additional decline in species and habitats. The assessment we describe adds value and novelty by evaluating both species and key habitats, setting a target reference point for biodiversity beyond no additional loss, integrating measures of social and ecological pressures that reduce biodiversity. These also help account for social and governance factors that should improve it.

This biodiversity assessment was developed as one of the ten goals that comprise the Ocean Health Index, with the objective of measuring how successfully the species and habitats that support biodiversity are being conserved. In this research study, the authors track both species and habitats in part due to limited data on marine species status. However, they also explicitly evaluate habitats because of the additional public values associated with maintaining a diverse set of marine environments. In addition, habitat condition can serve as a proxy indicator of species status for species that depend on habitat structure, but are not assessed. More than 90% of marine species have not yet been described and even fewer have been formally assessed for their status. Because species and habitat data each have gaps and shortcomings, by using data on both species and habitats, the authors were able to create an integrated, complementary measure of biodiversity that makes the best use of available data. They used this in order to explore in greater detail the nature and implications of the biodiversity results presented in the global Ocean Health Index, focusing in particular on country-by-country results. Results from their analyses reveal geographic, political, and governance patterns that may help explain successes and failures in biodiversity conservation. Through the results, they were also able to assess how specific habitat types and taxonomic groups affect Index scores. Finally, the authors were able to highlight where strategic action is likely to best promote biodiversity, where key data gaps remain, and how different assumptions and values affect our assessment and understanding of the current and future condition of biodiversity.

Threatened and Endangered Subspecies with Vulnerable Ecological Traits also have High Susceptibility to Sea Level Rise and Habitat Fragmentation

Assessing the vulnerability and adaptive capacity of species helps identify the most important factors affecting species survival and prioritize limited conservation resources. In addition to pressures such as habitat fragmentation invasive species disease, and overexploitation, climate is a critical factor affecting the survival and distribution of species. Climate affects biodiversity at multiple spatial and taxonomic scales, and recent climate change is already linked to alterations in species phenology, survival, and distribution. Climate change may also present unique challenges to threatened and endangered taxa because these species often exist in small populations, have limited genetic variation, and may have ecological traits that make them vulnerable to rapid environmental change. It is cost-effective to prioritize conservation efforts towards species and communities that have a relatively high probability of conservation success, and provide ecological, evolutionary, social, or economic value.

Florida is a national hotspot of endemism making biological conservation significant at local, national, and broader levels. There are also many taxa at risk of extinction, and only three other states in the country, Hawaii, California, and Alabama have more federally endangered taxa than Florida. Of the 572 endangered animal taxa listed on the federal register in the United States, 53 occur in Florida. Of these 53 federally listed taxa, 23 are designated as subspecies, and 19 of the 23 subspecies are endemic and unique to Florida. Understanding the vulnerability of endangered subspecies in this region to factors such as climate change and habitat fragmentation may help reduce extinction rates for Florida's endangered vertebrates.

Numerous pressures affect native species persistence in Florida, including climate change. Florida is particularly vulnerable to sea level rise, with about 10% of its land area less than 1 m above sea level. Most of the Florida Keys are predicted to be inundated or drastically altered under a sea level rise scenario of 0.6 m, and substantial tracts of the Everglades, other low-lying coastal areas, and barrier islands across the state will be submerged or severely modified with 1 m of sea level rise. Many areas that are not inundated by increased sea level will be susceptible to storm surges, flooding, erosion, and other risks. Although Florida's climate is predicted to warm less than other regions in North America, a climate inventory over the past 35 to 108 years indicated Florida is experiencing greater climate extremes, with trends of increased summer and fall maximum temperatures and decreased winter and spring minimum temperatures. The intensity of tropical storms is also predicted to increase, although frequency may decrease. Given the link between extreme climate events and the decline of local populations, an increase in frequency or intensity of extreme climate may threaten endangered species that exist in small fragmented populations.

High human population growth rates and non-native species invasions also threaten native fauna of this region. Florida's human population growth rate was the third highest in the United States between 2000 and 2010. Florida also has a high incidence of non-native plant and animal species. These pressures coupled with high biodiversity and limited conservation resources make setting conservation priorities extremely important.

Vulnerability assessments are a valuable tool that aid in prioritizing conservation efforts through the evaluation of threats and their impacts on species and communities. The Standardized Index for Vulnerability and Value Assessment (SIVVA) is a novel tool that is useful for evaluating vulnerability, adaptive capacity, and conservation value of species and communities. It provides several improvements over previous vulnerability assessments, such as the explicit incorporation of sea level rise and the ability to account for uncertainty in the assessment process. It allows for inclusion of the ecological, evolutionary, and economic value of taxa, and offers a flexible and model-like approach applicable to a broad range of taxonomic groups, including terrestrial, freshwater aquatic, and marine species. SIVVA allows for the evaluation of multiple interacting threats, both independently and collectively, to measure vulnerability and adaptive capacity, allowing for a more comprehensive assessment that is particularly useful for setting conservation priorities in regions that face numerous threats to biodiversity.

In this study the authors incorporated a multi-step approach to evaluate vulnerability and adaptive capacity of federally endangered subspecies in Florida. They specifically focused on subspecies pairs that differed in endangered status, because their phylogenetic relatedness provided the opportunity to ask questions about ecological traits related to adaptation and vulnerability. Because each subspecies in the taxon pair represented a closely related but distinct population, potential variation in ecological traits may help explain the relative ability to respond to environmental change. This framework allowed for a robust test to examine potential differences in traits for endangered taxa, especially given that controlled experiments on these subspecies are not feasible. Although comparing traits among related taxa and accounting for phylogenetic similarity was not uncommon in ecological studies, they were unaware of other regional-scale studies designed to investigate trait differences among many pairs of closely related taxa (subspecies) that varied in conservation status. The target group for this study included all of the federally endangered subspecies located in Florida.

In this study, they examined reproductive output, home range size, dispersal ability, and survival because they can affect population persistence and extinction risk. Their approach coupled SIVVA with in-depth literature surveys of ecological traits to evaluate vulnerability and adaptive capacity of endangered subspecies under future scenarios of sea level rise, habitat fragmentation, and climate change. Because endangered taxa often have restricted ranges, small population sizes, and reduced genetic variation, it was expected that endangered subspecies would demonstrate higher vulnerability and lower adaptive capacity compared to closely related, non-endangered subspecies. Four specific a priori hypotheses related to vulnerability criteria important in Florida were also developed. These predicted that endangered subspecies would have greater vulnerability to sea level rise, habitat fragmentation, and altered temperature and precipitation regimes. Lastly, the authors expected that endangered subspecies would have higher vulnerability and decreased adaptive capacity because of lower reproductive output, greater home range size, shorter dispersal, and lower survival rates.

Results suggested that the high threat of habitat inundation from sea level rise for endangered subspecies is of primary concern because approximately 10% of Florida's land area lies less than 1 m above current sea level. The mean percentage of habitat loss under a 1 m sea level rise scenario was 52% for endangered subspecies, compared to 11% for non-endangered subspecies. Additionally, all but one endangered subspecies showed high vulnerability to habitat fragmentation. Sea level rise vulnerability coupled with landscape fragmentation may limit habitat availability and inhibit dispersal into new areas, and highlights the challenge of managing human land use activities in conjunction with biodiversity conservation.

Evaluating the Combined Threat of Climate Change and Biological Invasions on Endangered Species

Climate change and invasive species are acknowledged as two of the most important causes of biodiversity loss in freshwater ecosystems, and they are expected to provoke major species extinctions in the near and long-term future. Under changing climatic conditions species are forced to adapt or to change their geographical range tracking climatic changes, and certainly many species migrations have already been documented in a variety of habitats. In comparison with native species, invasive species are more likely to adapt to the new climatic conditions because they are usually abundant, tolerate a broad range of climatic conditions, cover wide geographic ranges and have highly competitive biological traits. This will intensify the joint threat posed by climate warming and invasive species on native populations. While major efforts have been devoted to the study of climate change and invasive species separately, their combined impacts on the native fauna have rarely been assessed. This lack of explicit combined analysis of risk makes it difficult for science to inform effective prevention of species invasion and conservation of native fauna and flora. Consequently, forecasting the simultaneous response of invasive and native species to climate changes is of vital importance to both increase our understanding about species dynamics and promote adaptive management of the areas at highest risk of invasion

In order to approach this issue, the authors investigated the potential response to climate change of two pairs of invasive-native species. They chose as representative two of the most harmful and widespread freshwater invasive species in Europe: the zebra mussel (*Dreissena polymorpha*) and the signal crayfish (*Pacifastacus leniusculus*). Both of them are included into the list of 100 worst invaders in Europe. Amongst the various species potentially affected by these two invaders the depressed river mussel (*Pseudanodonta complanata*) and the white-clawed crayfish (*Austropotamobius pallipes*), were selected, both of which are included in the IUCN Red List of Threatened species.

While studies exist on the impacts of invasive species such as *D. polymorpha* and *P. leniusculus* on native populations, no study to date has explored the potential shift in these species' geographic ranges related to climate changes at a European scale. Considering their broad thermal tolerance and invasion ability, the authors predicted three separate hypotheses. The first one predicted the two invasive species to maintain or increase their range of distribution, most probably towards northern latitudes (hypothesis 1). On the other hand, they also predicted that the two native species are likely to present a contraction in their geographical niche due to their generally lower phenotypic plasticity (hypothesis 2). Finally they predicted, although not necessarily, such combination may result in greater overlap between both species further threatening the conservation of the native species (hypothesis 3). This study assumed that the modelled range of distribution of species is proportional to their realized niche and therefore changes in the modelled range (such as increases, decreases or range shifts) are likely to be reflected in the realized niche of the species. They also assumed that where both species overlap the invasive species will have a greater competitive advantage, hindering if not preventing completely the setting and spread of the native species. In order to test these predictions, ensemble Species Distribution Models (SDMs) were calibrated with each of the species' current distribution in Europe and projected onto present and future scenarios

(2050 A and B scenario families). The ultimate aim of this study was to support ecosystem management by anticipating future shifts in the distribution of native and invasive freshwater species over large spatial and temporal scales.

Results suggested that species distribution models displayed high performance for all four species and identified temperature-related variables as the most important predictors of their European distribution. Temperature has indeed a dominating effect on the life history, reproduction and growth of mussels, and has been identified as a major factor influencing crayfishes in terms of growth, mating and recruitment; survival, metabolic rate and reproduction; movement and dispersal. Other laboratory experiments have shown that lethal temperatures for both *P. leniusculus* and *A. pallipes* depend on the acclimation period, which is probably also true for other freshwater species including *D. polymorpha* and *P. complanata*. This may explain the discrepancies found in the literature about the adaptation of *P. leniusculus* to cold and warm climates, and suggests rapid temperature changes negatively affect both crayfishes, which is congruent with the high permutation importance the authors found of temperature seasonality in SDM.

Biotic Homogenization as a Threat to Native Affiliate Species: Fish Introductions Dilute Freshwater Mussel's Host Resources

The gradual increase in biological similarity of regions is a widespread process that shapes the composition and function of biotic communities and is mainly driven by the combined effects of species invasions and extinctions. As a result of biotic homogenization, many species begin interacting with novel partners, and former co-evolutionarily balanced inter-specific relationships are lost. The outcomes of these novel interactions determine the conditions for the survival of a particular species and have become one of the critical issues in conservation biology. Numerous studies have documented the direct negative effects of invading species on local biota via predation, competition or parasitism that often lead to other species becoming endangered. These direct impacts have been used in numerous models that demonstrate the threat of species introductions to global biodiversity. Less obvious consequences of biotic homogenization remain poorly understood but may be even more detrimental to local biota. Specifically, the cascading effects of species decline or extinction on another species across trophic levels may multiply the impacts on local biodiversity. Indeed, the loss of one species as a result of the loss of another species (co-extinction) is one of the most common causes of biodiversity loss.

Affiliate species, which directly depend on the presence of another species, are particularly threatened by biotic homogenization. Their ability to survive and prosper within a rapidly changing host community depends mainly on the broadness of a suitable host spectrum (host specificity) or on the capacity of the affiliate species to substitute its former hosts with incomers. Hence, the most threatened affiliates are considered to be the species that narrowly specialize on a few or even only one host species. Many examples from both the animal and plant kingdoms (e.g. insect parasites) document the decline or extinction of highly specialized species following the displacement of their exclusive hosts. Furthermore, hosts associated with many obligate dependent affiliate species may be considered 'keystone mutualists' with large conservation importance.

Although co-extinction rates and declines of highly specialized affiliate species have been well documented, the effects of biotic homogenization on affiliate species that are considered to be generalists remain understudied. It can be supposed that generalists may either use the remaining native species or the introduced species to compensate for the decline of their former hosts. Nevertheless, this assumes that the 'winners' and 'losers' of biotic homogenization are equally suitable hosts for an affiliate species. In contrast, recent evidence suggests that this assumption may not necessarily be true. Successful invaders are less parasitized in their invaded range compared with native species. Expanding species are often liberated from their ancestral affiliates and novel affiliates are not yet adapted to utilize the incomers. The role of biotic homogenization on the generalist affiliate species thus remains unclear.

Freshwater mussels (Bivalvia, Unionoida) have a short-term larval stage that is obligatory parasitic on the gills or fins of fishes. The availability of experimental methods for studying host compatibilities has resulted in the use of unionid bivalves as a common model group for the study of host–affiliate relationships involving endangered species. From a conservation point of view, unionids are among the most critically threatened groups of animals world-wide. The consequences of this catastrophic decline go far beyond the loss of species. This is to say that freshwater mussels have critical trophic and non-trophic functional roles in aquatic environment; thus, the decline of originally dense mussel populations can have interconnected implications for the functioning of aquatic ecosystems.

In this study, the authors examined the ability of a native affiliate species to exploit its host community despite the influx of non-native species. They tested the ability of non-native species to serve as alternative partners in local host–affiliate relationships. The European freshwater mussel *A. anatina* was used in this experiment, which is considered to be broad host generalist of native fish species and compared the compatibility of its glochidia with native versus non-native fishes in two distinct European regions. The studied regions were located in the central and peripheral part of the mussel's range to record the host–affiliate relationships within a broader biogeographical context and to investigate a possible variation in scenarios of host use throughout the species range. The authors then projected the obtained immunological host compatibility data into the recent progress of biotic homogenization and estimated the degree of host dilution. Finally, they discussed their outcomes in the context of strategies for conservation of endangered affiliate species.

From their results, the authors found significant differences in the ability of A. *anatina glochidia* to parasitize native and non-native fish species. As a result, the increasing presence of non-native species within fish communities across Europe likely decreases the availability of the mussel hosts. Hence, using the example of a native freshwater mussel, the authors were able to show that biotic homogenization of host communities may negatively interact with general life history traits such as host specificity of local affiliate species.

The Nature of Threat Category Changes in Three Mediterranean Biodiversity Hotspots

Red listing is a fundamental strategy for species conservation. Originally applied as a method to establish game control for hunting industries in Africa, it became a successful tool in threat analysis for vertebrates during the 1960s, and later was extended to plants. Red listing is now applicable to the whole compendium of biota on Earth and enables conservation practitioners to rank threatened species according to their threat status. Specifically for plants, a Red List is a floristic catalogue where plants are classified according to their level of threat. Usually these Red Lists are produced utilizing the expert opinion of the scientific community and any quantitative data available for the species. They discuss and evaluate conservation problems and risks of those species considered to be threatened, and test known data against criteria thresholds, to arrive at a consensus on conservation status and the Red List category of threat for each plant species.

The most wide-spread classification system has been developed under the auspices of the International Union for the Conservation of Nature (IUCN) since the last century. Currently, the IUCN system classifies threatened species into threat categories (critically endangered, endangered, and vulnerable) following qualitative and quantitative criteria based on range and population size, condition and demographic trends. A Red List summarizes the conservation knowledge of the species at a particular time, so it is important to produce Red List assessments on a regular basis to identify changes in the status of the biota and specific conservation trends.

Regular assessments are one of the few tools to show the changing nature of conservation status across time, with important advances being made in such assessments for some taxonomical groups. However, little is known about the mobility of plants across threat categories; that is, how fast and how far do plants move along the conservation category continuum from one Red List to the next one to produced.

There are several factors contributing to changes in the conservation status of plants. Plants step down or up in a Red List for reasons related to conservation matters (worse or better scenarios) but not exclusively. For example, the work of taxonomists lumping or splitting taxa may produce down or upgrading respectively.

In addition, improvement in knowledge of the taxon relating to its conservation status is also a factor affecting movement across categories. Finally, classification systems are not immutable, they evolve according to the improvement in knowledge in conservation biology, whether in new methods or in new concepts. Therefore changes in category definitions and limits as well as criteria used for the category assignment, have been and will be in constant change following progress in plant conservation science. Those reasons shape the movement of plants across conservation categories.

To explore this idea the authors proposed their first null hypothesis as "changes in categories have no sensitivity to the reason or factor dominance". If the null hypothesis was rejected they may find that some changes may be more affected for a particular factor than for others. For example, changes in taxonomy may produce a concentration of critically endangered species due to taxonomic splitting, or, changes in knowledge may produce a high amount of species moving to lower threat categories because more populations or individual are discovered.

Reporting threats in plants is closely associated with Red listing and conservation status assessments. For that, it is necessary to know which are the most important threats for a particular biota; how these threats are affecting the threatened taxa across their area range; and what emerging patterns may arise. One may expect that causes or origin of threats are highly idiosyncratic depending on location. They also may depend on the quality of information which in turn depends on intensity of field surveys and the level of taxonomic knowledge, among other things.

In spite of these issues it is possible that some types of threats may be associated with particular step changes in categories. For example, threats associated with land use changes are more prone to produce real conservation upgrading of threatened taxa than other types of threats that are more difficult to quantify, such as those related to stochastic environmental changes or intrinsic biological problems such as lack of pollinators or natural competition. This is especially the case when monitoring spans relatively short time frames compared to the generation time of the taxon. This led the authors to create a second null hypothesis which states: "threat type is independent of step changes in threat category".

To test these two hypotheses, the authors used data from three Mediterranean-type ecosystems (MTEs): California, Spain and Western Australia. MTEs share particular distinct evolutionary conditions, and are one of the most densely human populated and diverse biomes in the world. MTEs tend to have well established threat processes such as land clearing, grazing or fire disruption which interact with novel risks such as introduced species and urban development. There are also a profusion and diversity of conservation resources able to be directed at managing threats and threatened species. The diversity of both conservation problems and resources available to combat them in MTE may enable researchers to draw some general conclusions to better manage biodiversity losses in other parts of the World. One tool available to assess the net effect of these factors is plant Red Lists, which have been systematically produced since the 1980s in some of these Mediterranean regions. Such plant Red Lists have been produced by scientific communities in California, Spain and Western Australia The authors used these in order to test whether their hypotheses were correct.

Results for this study showed that Red Plant Lists are dynamic, with plant threat classifications changing temporally. The reasons for these changes are varied and not necessarily linked to the types of threat processes affecting the species. Most of the observed changes in threat category were determined by causes not directly related to natural conservation changes, with increasing knowledge about a species from one listing date to the next being the most frequent cause of change in all studied territories. It also showed that when a threat is reported for Western Australian, Spanish or Californian plants, there is a 0.83 chance that this threat will be related to human pressures. Finally results suggested that plants that remained in the same conservation status from 2000 to 2008 were less frequently affected by threats related to humans, predictable and involving space loss than the average plant.

References Cited

Aslan, C. E., Zavaleta, E. S., Tershy, B., Croll, D., 2013. Mutualism Disruption Threatens Global Plant Biodiversity: A Systematic Review. DOI: 10.1371/journal.pone.0066993.

Benscoter, A. M., Reece, J. S., Noss, R. F., Brandt, L. A., Mazzotti, F. J., Romanach, S.S., Watling, J. I., 2013.Threatened and Endangered Subspecies with Vulnerable Ecological Traits Also Have High Susceptibility to Sea Level Rise and Habitat Fragmentation. DOI: 10.1371/journal.pone.0070647.

Douda, K., Lopes-Lima, M., Hinzmann, M., Machado, J., Varandas, S., Teixeira, A., Sousa, R. 2013. Biotic homogenization as a threat to native affiliate species: fish introductions dilute freshwater mussel's host resources. DOI: 10.1111/ddi.12044.

Fuentes, M. M. P. B., Pike, D. A., Dimatteo, A., Wallace, B. P, 2013. Resilience of marine turtle regional management units to climate change. DOI: 10.1111/gcb.12138.

About the Authors

The authors of this book are students at the Claremont Colleges. The book is a work product of Biology 159: Natural Resources Management taught by Emil Morhardt in the Joint Science Department of Claremont McKenna, Pitzer, and Scripps Colleges. Each student picked a topic, did a full literature search, and selected eight papers written within the past year that exemplified the state of the science.

Their task was to write journalistic summaries capturing the essence of the papers but eschewing technical terms to the extent possible—to become, in effect, science writers. The summaries were due weekly and were returned with editorial comments shortly thereafter. The chapters are compilations of the individual summaries with additional introductory and conclusionary material.

The editor is Roberts Professor of Environmental Biology at Claremont McKenna, Pitzer, and Scripps colleges, and Director of The Roberts Environmental Center at Claremont McKenna College. He remembers how difficult it is to learn to write and appreciates the professionalism shown by these students.

Index

www.ingramcontent.com/pod-product-compliance
Lightning Source LLC
Chambersburg PA
CBHW031501270326
41930CB00006B/199

* 9 7 8 0 9 8 4 3 8 2 3 8 5 *